Alternative
Energy
Demystified

Demystified Series

Accounting Demystified
Advanced Statistics Demystified
Algebra Demystified
Alternative Energy Demystified
Anatomy Demystified
asp.net 2.0 Demystified
Astronomy Demystified
Audio Demystified
Biology Demystified
Biotechnology Demystified
Business Calculus Demystified
Business Math Demystified
Business Statistics Demystified
C++ Demystified
Calculus Demystified
Chemistry Demystified
College Algebra Demystified
Corporate Finance Demystified
Databases Demystified
Data Structures Demystified
Differential Equations Demystified
Digital Electronics Demystified
Earth Science Demystified
Electricity Demystified
Electronics Demystified
Environmental Science Demystified
Everyday Math Demystified
Forensics Demystified
Genetics Demystified
Geometry Demystified
Home Networking Demystified
Investing Demystified
Java Demystified
JavaScript Demystified
Linear Algebra Demystified
Macroeconomics Demystified
Management Accounting Demystified

Math Proofs Demystified
Math Word Problems Demystified
Medical Billing and Coding Demystified
Medical Terminology Demystified
Meteorology Demystified
Microbiology Demystified
Microeconomics Demystified
Nanotechnology Demystified
Nurse Management Demystified
OOP Demystified
Options Demystified
Organic Chemistry Demystified
Personal Computing Demystified
Pharmacology Demystified
Physics Demystified
Physiology Demystified
Pre-Algebra Demystified
Precalculus Demystified
Probability Demystified
Project Management Demystified
Psychology Demystified
Quality Management Demystified
Quantum Mechanics Demystified
Relativity Demystified
Robotics Demystified
Signals and Systems Demystified
Six Sigma Demystified
SQL Demystified
Statics and Dynamics Demystified
Statistics Demystified
Technical Math Demystified
Trigonometry Demystified
UML Demystified
Visual Basic 2005 Demystified
Visual C# 2005 Demystified
XML Demystified

Alternative
Energy
Demystified

Stan Gibilisco

New York Chicago San Francisco Lisbon London
Madrid Mexico City Milan New Delhi San Juan
Seoul Singapore Sydney Toronto

The McGraw·Hill Companies

McGraw-Hill books are available at special quantity discounts to use as premiums and sales promotions, or for use in corporate training programs. For more information, please write to the Director of Special Sales, Professional Publishing, McGraw-Hill, Two Penn Plaza, New York, NY 10121-2298. Or contact your local bookstore.

Alternative Energy Demystified

1 2 3 4 5 6 7 8 9 0   DOC DOC   0 1 9 8 7 6

ISBN-13: 978-0-07-147554-9
ISBN-10:   0-07-147554-0

Sponsoring Editor
Judy Bass

Editorial Supervisor
Janet Walden

Project Manager
Patricia Wallenburg

Copy Editor
William McManus

Proofreader
Paul Tyler

Indexer
Stan Gibilisco

Production Supervisor
Jean Bodeaux

Composition
Patricia Wallenburg

Illustration
Patricia Wallenburg

Cover Series Design
Margaret Webster-Shapiro

Cover Illustration
Lance Lekander

To Samuel, Tim, Tony, and Remy

ABOUT THE AUTHOR

Stan Gibilisco is one of McGraw-Hill's most prolific and popular authors. His clear, reader-friendly writing style makes his books accessible to a wide audience, and his experience as an electronics engineer, researcher, and mathematician makes him an ideal editor for reference books and tutorials. Stan has authored several titles for the McGraw-Hill *Demystified* library of home-schooling and self-teaching volumes, along with more than 30 other books and dozens of magazine articles. His work has been published in several languages. *Booklist* named his *McGraw-Hill Encyclopedia of Personal Computing* one of the "Best References of 1996," and named his *Encyclopedia of Electronics* one of the "Best References of the 1980s."

CONTENTS

PREFACE

This book is for people who want to learn about diverse energy sources and technologies without taking a formal course. It can serve as a classroom supplement, tutorial aid, self-teaching guide, or home-schooling text.

As you take this course, you'll encounter multiple-choice quizzes and a final exam to help you measure your progress. All quiz and exam questions are composed like those in standardized tests. The quizzes are "open-book." You may refer to the chapter text when taking them. The final exam contains questions drawn uniformly from all the chapters. It is a "closed-book" test. Don't look back at the text when taking it. Answers to all quiz and exam questions are listed at the back of the book.

You don't need a mathematical or scientific background for this course. Middle-school algebra, geometry, and physics will suffice. I recommend that you complete one chapter a week. That way, in a few months, you'll finish the course. You can then use this book, with its comprehensive index, as a permanent reference.

This book offers ideas for consumers, experimenters, and hobbyists, as well as outlining the technical basics of energy generation, transport, and utilization. However, this is not a design guide! If you want to install, modify, upgrade, or use any of the systems discussed here, consult the appropriate professionals, and adhere to all applicable laws, codes, and insurance requirements.

This is an entry-level science nonfiction book for students and lay people. It is not intended to promote or condemn any particular energy source, ideology, agenda, or economic interest. I have done my best to objectively present the advantages and limitations of various technologies from conventional to exotic. I invite input from innovators, producers, and distributors concerning developments for possible inclusion in future editions.

Stan Gibilisco

CHAPTER 1

Heating with Wood, Corn, and Coal

For some people, burning dead plant matter or solid fossil fuels can make the difference between comfort and freezing. For others, such fuels are cheaper or more easily available than conventional heat sources such as methane, propane, or oil. Let's look at some "primitive" but time-proven methods of home heating. The generic systems described here are typical, but variations abound.

Energy, Power, and Heat

Have you heard the terms *energy*, *power*, and *heat* used interchangeably as if they mean the same thing? They don't! Energy is power manifested over time. Power is the rate at which energy is expended. Heat is any form of energy transfer that causes changes in temperature. Energy, power, and heat can be expressed in several ways, and can occur in various forms.

THE JOULE

Physicists measure and express energy, regardless of its form, in units called *joules*. One joule (1 J) is the equivalent of one *watt* (1 W) of power expended, radiated, or dissipated for one *second* (1 s) of time. A joule is the equivalent of a *watt-second*, and a watt is the equivalent of a *joule per second*. Mathematically:

$$1 \text{ J} = 1 \text{ W} \cdot \text{s}$$
$$1 \text{ W} = 1 \text{ J/s}$$

In electrical heating systems, you'll encounter the *watt-hour* (symbolized W · h or Wh) or the *kilowatt-hour* (symbolized kW · h or kWh). A watt-hour is the equivalent of 1 W dissipated for 1 h, and 1 kWh is the equivalent of one *kilowatt* (1 kW) of power dissipated for 1 h. Note that 1 kW = 1000 W. Therefore:

$$1 \text{ Wh} = 3600 \text{ J}$$
$$1 \text{ kWh} = 3{,}600{,}000 \text{ J} = 3.6 \times 10^6 \text{ J}$$

THE CALORIE

A less often used unit of heat is the *calorie*. One calorie (1 cal) is the amount of energy transfer that raises the temperature of exactly one gram (1 g) of pure liquid water by exactly one degree Celsius (1°C). It is also the amount of energy lost by 1 g of pure liquid water if its temperature falls by 1°C. The *kilocalorie* (kcal), also called a *diet calorie*, is the amount of energy transfer involved when the temperature of exactly one kilogram (1 kg), or 1000 g, of pure liquid water rises or falls by exactly 1°C. It turns out that 1 cal = 4.184 J, and 1 kcal = 4184 J.

This definition of the calorie holds true only as long as the water is liquid during the entire process. If any of the water freezes, thaws, boils, or condenses, this definition is not valid. At standard atmospheric pressure on the earth's surface, in general, this definition holds true for temperatures between approximately 0°C (the freezing point of water) and 100°C (the boiling point).

THE BRITISH THERMAL UNIT (BTU)

In home heating applications in the United States, an archaic unit of energy is used: the *British thermal unit* (Btu). You'll hear this unit mentioned in advertisements for furnaces and air conditioners.

One British thermal unit (1 Btu) is the amount of energy transfer that raises the temperature of exactly one pound (1 lb) of pure liquid water by exactly one degree Fahrenheit (1°F). It is also the amount of energy lost by 1 lb of pure liquid water if its temperature falls by 1°F. This definition, like that of the calorie, holds true only as long as the water remains in the liquid state during the entire process.

If someone talks about "Btus" literally, in regards to the heating or cooling capacity of a furnace or air conditioner, that's an improper use of the term. They really mean to quote the rate of energy transfer in *British thermal units per hour* (Btu/h), not the total amount of energy transfer in British thermal units. The real-world heating ability of a stove or furnace is expressed in terms of power, not energy. As things work out, 1 Btu = 1055 J. Another, more useful, pair of facts are these:

$$1 \text{ Btu/h} = 0.293 \text{ W}$$
$$1000 \text{ Btu/h} = 293 \text{ W}$$

Conversely:

$$1 \text{ W} = 3.41 \text{ Btu/h}$$
$$1 \text{ kW} = 3410 \text{ Btu/h}$$

A home furnace with a heating capacity of 100,000 Btu/h operates at the equivalent of 29.3 kW. That's roughly the amount of power consumed by 20 portable electric space heaters operating at "full blast."

FORMS OF HEAT

If you place a kettle of water on a hot stove, heat is transferred from the burner to the water. This is *conductive heat*, also called *conduction* (see Figure 1-1, part A). When an *infrared* (IR) lamp shines on your sore shoulder, energy is transferred to your skin surface from the filament of the lamp. This is *radiative heat*, also called *radiation* (see Figure 1-1, part B). When a fan-type electric heater warms a room, air passes through the heating elements and is blown into the room, where the heated air rises and mixes with the rest of the air in the room. This is *convective heat*, also called *convection* (see Figure 1-1, part C).

The Wood Stove

Wood fuel is the oldest way to obtain artificial heat. *Wood stoves* have become sophisticated in recent years, with the advent of optimized air intake systems, blowers, thermostats, and *catalytic converters* similar to the emission-control devices found in cars.

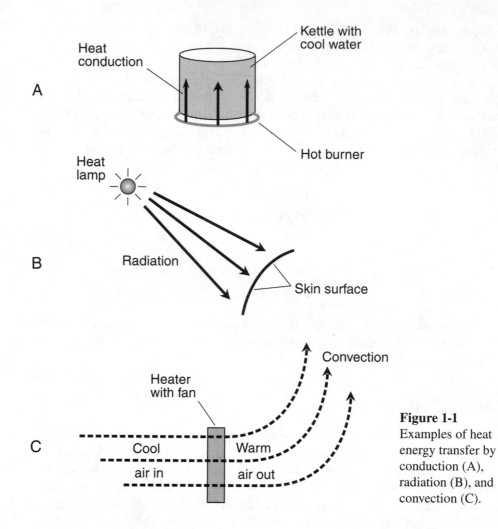

Figure 1-1
Examples of heat energy transfer by conduction (A), radiation (B), and convection (C).

HOW IT WORKS

In a wood stove, a controlled fire heats a heavy cast-iron box, which in turn emits heat in the form of radiation. This IR energy warms the walls, floor, ceiling, and furniture. In addition, heat is transferred to the air by conduction: direct contact with the hot stove and with the warmed walls, floor, ceiling, and furniture. The warmed air rises, causing continuous air circulation (convection) that helps to equalize the temperature throughout the room. A wood stove therefore heats a room by all three modes familiar to the physicist (see Figure 1-2).

A good wood stove can heat a large room in a reasonable amount of time, even when the outside temperature is far below freezing. If the stove is installed in the basement of a house near the main air intake vent for a conventional furnace, the

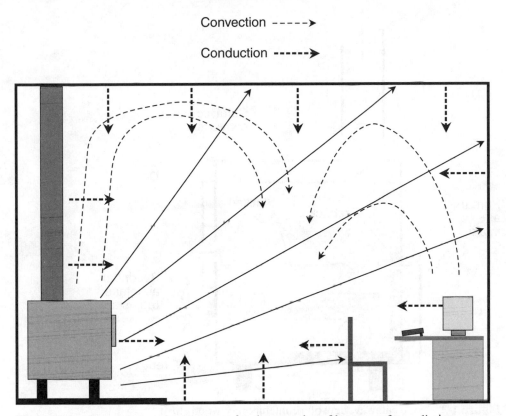

Radiation ⟶
Convection ----➤
Conduction ----➤

Figure 1-2 A wood stove heats a room by three modes of heat transfer: radiation, convection, and conduction.

furnace blower can circulate the heated air throughout the house even if the furnace itself is not operating. In this way, a large wood stove can keep a medium-sized house warm in all but the coldest weather.

Figure 1-3 is a cutaway view of a typical wood stove as seen directly from the side. The *primary air intake* ensures that some air always flows into the *firebox*. The intensity of the fire can be controlled by adjusting the *secondary air intake*. Opening this valve increases the rate of the burn and increases the temperature. Closing it reduces the burn rate to a minimum. The *catalytic converter* changes most of the energy contained in the smoke into usable heat, and also reduces the particulate pollution that goes up the *stack*. (The catalytic converter can be bypassed if desired.) A large wood stove can provide upwards of 150,000 Btu/h of heating power, provided it is kept operating properly. This is comparable to a large gas furnace.

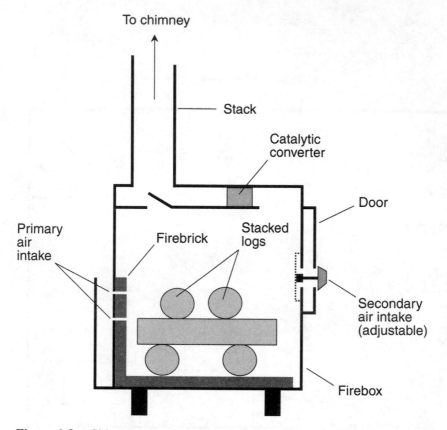

Figure 1-3   Side cutaway view of a contemporary wood stove.

An excellent reference for wood stove operation, as well as wood fuel in general, is a little book called *All That's Practical About Wood* by Ralph W. Ritchie (Springfield, Oregon: Ritchie Unlimited Publications, 1998). However, neither that book nor this one is intended to serve as a safety guide. If you have any doubt about the installation and use of a wood stove after reading its instruction manual, contact your local fire marshal, who will want to inspect the system anyway.

ADVANTAGES OF THE WOOD STOVE

- The wood stove requires no external power source to heat the room in which it is located. If that room is on the lower level, doors can be left open so the warm air will rise to heat the rest of the house. In this way, the wood stove can serve as an emergency heat source when all normal utilities have gone down.

- Wood is a renewable fuel. Trees can be deliberately grown and harvested to provide fuel for heating, just as trees are grown and harvested to provide lumber for building.

- Burning wood in a stove can minimize waste. Wood that would otherwise be burned at a brush dump or create a wildfire hazard (dead wood in a forest, for example) can be gathered, cut, and used to heat homes.

- Frequent and regular use of a wood stove can significantly reduce the cost of heating a home by conventional means. It can also mitigate the impact of a sudden, severe shortage of natural gas or oil.

- For some people, wood stoves have esthetic appeal.

LIMITATIONS OF THE WOOD STOVE

- Wood stoves can be dangerous! Before installing and using one, read the instruction manual. You should wear safety glasses and completely cover yourself (including your hands) with fireproof clothing when working around an active wood stove.

- You must acquire and maintain a stockpile of dry, cut wood. Logs must be split and cut to lengths small enough to easily fit in the firebox. This can be inconvenient. In some locations, cut wood is extremely expensive.

- Wood must be allowed to dry for at least 12 months after being cut, and preferably for 18 months. Fresh-cut wood has high moisture content, and such wood burns inefficiently (and sometimes won't burn at all).

- The fire requires constant attention.

- Wood is relatively inefficient as a fuel source. No wood stove can equal the efficiency of a top-of-the-line gas furnace.

- A wood stove requires frequent cleaning. Ashes and coals must be allowed to completely cool before removal, and this translates into stove downtime.

- The chimney needs periodic cleaning to prevent buildup of *creosote*, which can ignite and cause a dangerous *flue fire* (also known as a *chimney fire*).

- Wood stoves are restricted or forbidden in some municipalities. Some insurance companies won't underwrite a policy for a home that has a wood stove.

- If you want to heat your whole house with a wood stove, the room where the stove is located will become extremely hot.

PROBLEM 1-1
What can be burned in a wood stove besides wood to provide heat? How about charcoal, or coal, or flammable liquids?

SOLUTION 1-1
Most wood stoves are designed to burn properly cut, dry wood, and nothing else. Charcoal or coal gets too hot. The use of any flammable liquid can cause an explosion and set clothes, carpeting, and furniture on fire instantly. A few specialized wood stoves are designed to burn coal as well as wood. This is discussed later in the chapter in the section "Coal Stoves."

Pellet Stoves and Furnaces

There's a more efficient, cleaner, and safer way to burn wood than the old-fashioned "logpile" method. Sawmills compress waste sawdust into pellets that can be burned in *pellet stoves* and *pellet furnaces*.

HOW THEY WORK

Figure 1-4 is a simplified functional diagram of a pellet stove. The pellets, which look a little like dry pet food, must be poured into a *hopper*. A feed system, usually consisting of a *motor* and *auger* or other mechanical device, supplies pellets to the firebox at a rate that can be set manually or automatically, depending on the type of stove and on user preference.

Wood pellets are too energy-dense to burn in a free-standing pile. You can't fill up an ordinary wood stove with pellets and expect it to work. In order for combustion to take place, air must be forced through the pellet pile. A pellet stove has a *blower* that forces air through the firebox, ensuring combustion. The air can be taken from outside the house to prevent negative pressure that would otherwise draw cold air into the house. The exhaust fumes are vented to the outside as well. Heated, unpolluted air from inside the stove, after having been warmed by the firebox and a corrugated mass of metal called a *heat exchanger*, is blown into the room.

A pellet furnace is basically an oversized pellet stove. The hot air is blown into the ductwork that circulates it throughout the house. Pellet furnaces can be installed directly in place of forced-air gas furnaces with little or no modification to the existing air distribution system. Free-standing pellet stoves are generally rated from 30,000 to 70,000 Btu/h. Pellet furnaces can deliver considerably more heat energy than that.

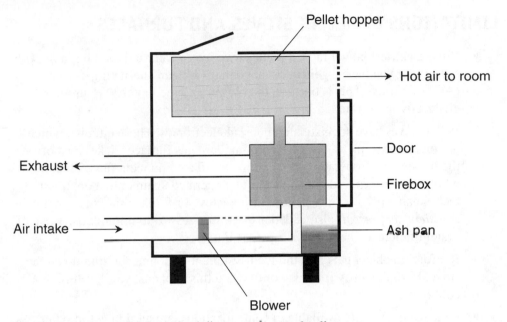

Figure 1-4 Simplified functional diagram of a wood pellet stove.

ADVANTAGES OF PELLET STOVES AND FURNACES

- Pellet stoves are more efficient than wood stoves. The pellets are refined, so they contain minimal moisture, little or no pitch (sap), no dirt, no insects, and no bark. The result is more heat, less pollution, and less ash per kilogram of fuel.

- Pellet stoves are safer than wood stoves. The exterior of the pellet stove does not become dangerously hot (except for the door glass).

- With a pellet stove, the temperature is easier to regulate than is the case with a wood stove. The pellet stove doesn't need constant attention. You can set a thermostat and pretty much forget it, except for periodic hopper refilling.

- The ash can be disposed of without extended periods of stove downtime.

- Pellet stoves do not require chimneys. The exhaust gases can be vented out the side of the house, in the same manner as is done with high-efficiency gas furnaces. Thus, there is no buildup of creosote.

- Pellet stoves or furnaces may be allowed in regions or municipalities where wood stoves are forbidden.

LIMITATIONS OF PELLET STOVES AND FURNACES

- If the electrical power fails, a pellet stove won't work unless it has a *backup battery*, or you have a generator that creates a clean alternating-current (AC) *sine wave*. This is because the blower and the motor(s) require electricity to operate.

- Pellet stoves have sophisticated internal electronics. These circuits, which are much the same as those found in modern gas furnaces, take most of the hassle out of operating the system—until a component fails. Then the whole machine goes down, and can't be operated again until a qualified technician repairs it. Any pellet-burning stove or furnace should have a *transient suppressor*, also called a *surge suppressor*, to minimize the risk of system failure as the result of a power-line "spike."

- If a foreign object gets into the feed system, it will jam, shutting down the stove. If you're away for a day or two and this happens, you'll return to a cold house.

- Pellets, while easily available in some locations, are hard to get in other places. You'll have to stockpile them, in much the same way as you stockpile wood for a wood stove.

- Pellets come in heavy bags, usually 18 kilograms, which is 40 pounds. In cold weather, the pellet hopper will have to be filled once a day or more. That means you'll be lifting and hauling a lot of those bags.

PROBLEM 1-2
Can a wood or pellet stove be safely vented into the same chimney as another appliance such as a gas furnace?

SOLUTION 1-2
A single chimney can have multiple flues (insulated, fireproof air ducts leading to the outside), each of which serves a different appliance. But such a chimney should be inspected by the fire marshal, and by your insurance company, before any of the appliances is used. *Every wood stove or furnace should be vented into a dedicated flue. You are courting trouble if you connect a wood- or pellet-burning system to the same flue as any other appliance.* Local codes are not always clear about this issue, but a little research on the Internet ought to convince you that if you want to remain physically healthy and financially solvent and live in an undamaged home, you had better not vent a wood or pellet stove into the same flue as any other appliance.

Corn Stoves and Furnaces

As an alternative to wood pellets, shelled dry corn can be burned in *corn stoves* and *corn furnaces*. These systems resemble pellet stoves and furnaces in many ways, but there are some important differences.

HOW THEY WORK

Figure 1-4, shown earlier, can serve as a simplified functional diagram of a corn stove. The most obvious difference between the corn stove and the pellet stove is the fact that, rather than wood pellets, individual corn kernels (cleaned of the cob, corn silk, and other foreign matter) are poured into the hopper. The feed system supplies the kernels to the firebox.

Corn stoves and furnaces differ from pellet systems in other ways, too. Corn contains significant amounts of *ethanol* (the same ethanol used in alternative fuels such as *gasohol* or *E85* for cars and trucks), whereas wood does not. Ethanol burns hotter than wood. In addition, corn contains oil (the same stuff you can use to fry food), which also burns hotter than wood, although more slowly than ethanol.

The waste matter in a corn stove accumulates in the form of a *clinker*, which is like a super-concentrated lump of coal. This clinker must be periodically removed and discarded, just as the ash from a pellet system must be discarded. The rest of the corn system is pretty much the same as the pellet system.

Corn stoves are rated from approximately 30,000 to 70,000 Btu/h. Corn furnaces, designed for the forced-air heating of entire homes, can deliver considerably more, in some cases upwards of 100,000 Btu/h.

ADVANTAGES OF CORN STOVES AND FURNACES

- Corn stoves are more efficient than cut-wood stoves. The corn kernels are dry and clean, and burn almost completely. The result is maximum heat, with minimum pollution and waste matter.

- Corn stoves are safer than cut-wood stoves, for the same reasons that apply to pellet stoves.

- With a corn heating system, the temperature is easy to regulate, just as is the case with a pellet system.

- The clinker can be disposed of without extended downtime, and makes less of a mess than the ash that results from the burning of cut wood or even wood pellets.

- Corn stoves and furnaces, like pellet systems, do not require chimneys, so all the problems associated with chimneys need be of no concern. The exhaust gases can be vented out the side of the house.
- Corn-based heating systems may be allowed in regions or municipalities where wood-burning systems are forbidden.
- If you're good friends with a farmer who produces a surplus of corn almost every year, you are doubly blessed!

LIMITATIONS OF CORN STOVES AND FURNACES

- If the electrical power fails, a corn stove won't work. It's the same problem that occurs with a pellet system. You can get a backup battery to run the auger and the blower, but this battery must be precharged, and it won't work for more than a few hours in the absence of AC power.
- Corn stoves and furnaces, like their pellet counterparts, contain electronic circuits that are susceptible to power-line surges. A corn-burning system should therefore employ a transient suppressor in the AC power line.
- If a foreign object gets into the feed system, you'll have the same trouble as you'll have if it happens in a pellet-burning stove or furnace.
- Refined, dried corn, while easily available in some locations, is impossible to get in other places, unless you have it shipped in at great expense.

PROBLEM 1-3
Don't the kernels in a corn stove or furnace sometimes pop uncontrollably, causing noise and a mess, and giving rise to the risk of an explosion?

SOLUTION 1-3
This won't happen in a properly operating system. In fact, the corn won't even snap or crackle. If the above-mentioned scenarios were a problem, corn stoves wouldn't have survived on the market.

Coal Stoves

In the United States, abundant coal reserves still exist. Because of this, and also because of increasing prices for conventional heating fuels and the continued exotic nature of more technologically advanced heating methods, coal-burning stoves have become popular in recent years.

HOW THEY WORK

Coal stoves resemble wood stoves. In fact, hybrid units exist that can burn either coal or wood. The main difference between a coal-burning stove and a wood-burning stove is in the nature of the air intake system. Wood burns best with air supplied mainly from above, while coal burns better when the air comes in from underneath.

Figure 1-5 is a functional diagram of a hybrid stove that can be used to burn either coal or wood. For wood burning, the coal air intake damper is closed, and the wood air intake is used to adjust the air flow and the rate of combustion. For coal burning, the wood air intake is closed and the coal air intake damper is opened. In

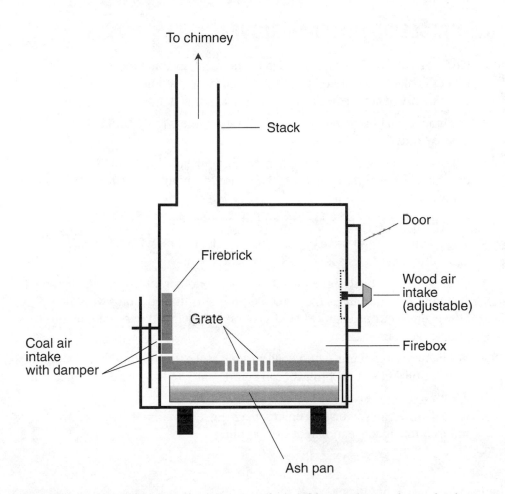

Figure 1-5 Simplified functional diagram of a hybrid stove that can be used to burn cut wood or anthracite coal.

either situation, the fuel produces ash that collects in a pan at the bottom of the stove. This pan must be periodically removed and emptied.

Some coal-burning stoves have thermostat-controlled dampers that regulate the air flow, and consequently the burn rate, when coal is used. A more advanced stove may have a blower at the coal air intake point. Coal-burning stove manufacturers recommend the burning of deep-mined *anthracite* coal. The use of *bituminous* coal is discouraged. *Peat* and *lignite* can be burned in some hybrid systems as if they were wood, but the instruction manual should be consulted concerning this.

Wood-only stoves should not be used to burn coal, because the air intake system is not designed for any fuel other than dry, cut wood. Similarly, coal-only stoves should not be used to burn anything other than deep-mined anthracite.

ADVANTAGES OF THE COAL STOVE

- Coal is easily available, and is also economical, in some locations. This makes coal a viable alternative energy source for people who live in certain places. (If you are one of these people, you know it!)
- Anthracite coal burns fairly clean, contrary to the widely held belief that all coal is "dirty."
- Coal stoves are efficient and, when properly installed with an air distribution system, can heat a small house to a comfortable temperature even in frigid weather.
- Coal does not have to be manufactured, as do pellets.
- The availability and price of coal does not depend on what happens in the agricultural market, as is the case with corn.
- Coal stoves can be designed without augers or other electromechanical feed systems. This eliminates the potential problems that go along with them. However, electromechanical feed systems are included in some units, for people who prefer them.
- Frequent and regular use of a coal-burning stove, in addition to a conventional gas or oil furnace, can significantly reduce the cost of heating a home.
- A coal stove can serve as an emergency heat system in the event of an interruption in conventional utility supplies.

LIMITATIONS OF THE COAL STOVE

- Coal stoves can be dangerous, for the same reasons wood stoves can be dangerous.

- A supply of coal must be maintained at the home site. This can be inconvenient, and can also be objectionable to some people.

- Although coal does not pollute as much as some people imagine, it is not as clean-burning as oil or natural gas.

- A coal fire requires a lot of attention, unless the stove can hold a large quantity of coal and has an automatic feed system.

- A coal stove requires frequent cleaning. The chimney also needs periodic cleaning and inspection to ensure that it remains in good overall condition.

- If you want to heat your whole house with a coal stove, the room where the stove is located will become extremely hot, unless an air-distribution system is employed.

- Coal is not easy to obtain in all locations.

- Some local or municipal governments restrict or prohibit the use of coal-burning stoves.

PROBLEM 1-4

I've read stories about the use of coal and wood stoves during severe winters of the Northern Great Plains in pioneer days. One story told of how the coal stove got so hot that it glowed. Isn't this dangerous?

SOLUTION 1-4

If a coal or wood stove gets so hot that it visibly glows, the combustion rate is too high. This can permanently ruin the stove, and might cause sudden structural failure of the firebox. Then you'll have live coals on the floor! If you think a coal or wood stove is getting too hot, turn it down by restricting the air intake.

PROBLEM 1-5

Suppose you have decided that you want to install a wood, pellet, corn, or coal stove to back up your conventional furnace. (You'll decide which type eventually.) You are concerned about fire safety. What should you do?

SOLUTION 1-5

Consult the people in your fire department. Consult the fire marshal. Pick their brains! Obtain all the pamphlets and other data you can from them, heed all local regulations, be sure your insurance company knows (and approves of) what you're doing, and get a final, official inspection of the system after installation. Have fire extinguishers handy, use smoke detectors and a *carbon monoxide* (CO) detector in your house, and devise an evacuation plan in case of the worst. Some towns and counties offer fire-safety seminars and classes. Take advantage of them.

Quiz

This is an "open book" quiz. You may refer to the text in this chapter. A good score is eight correct. Answers are in the back of the book.

1. Which of the following units can be used to express heat power?

    a. The British thermal unit per hour (Btu/h)

    b. The joule (J)

    c. The calorie (cal)

    d. The kilowatt-hour (kWh)

2. Which of the following is a disadvantage of corn as a fuel source?

    a. It does not burn efficiently.

    b. It contains ethanol, which can cause explosions under certain circumstances.

    c. It contains oil that evaporates and then re-condenses as creosote.

    d. It may not be readily available.

3. A stove designed to burn cut, dried wood operates best when

    a. corn, pellets, or coal are used in addition to the wood.

    b. the logs are exposed to air intake from above.

    c. the logs are exposed to air blown through from underneath.

    d. the logs are fresh-cut, so they won't burn too fast.

4. Imagine that you have built a one-room cabin. You heat it with a wood stove. You place the stove in the middle of the room, with a stack that goes straight up and out through the roof. When the stove operates, warm air rises in the middle of the room, flows outward along the ceiling, and descends down along the walls; then it flows inward along the floor toward the stove, where it is heated and rises again. This is an example of how warm air is distributed throughout a room by means of

    a. radiation.

    b. conduction.

    c. convection.

    d. precipitation.

5. The best type of fuel for burning in a coal-only stove is

    a. peat.

    b. lignite.

    c. bituminous.

    d. anthracite.

6. Which of the following types of heating system can operate indefinitely if the AC electrical power fails, even without a backup generator or battery?

    a. A stove designed to burn shelled, dried corn

    b. A furnace designed to burn wood pellets

    c. A stove designed to burn cut wood

    d. All of the above

7. A kilowatt is the equivalent of

    a. 1 J/s.

    b. 1000 J/s.

    c. $1 \text{ J} \cdot \text{s}$.

    d. $0.001 \text{ J} \cdot \text{s}$.

8. Suppose an electric heater operates at exactly 800 W. Assuming all the electricity is converted to usable heat, what is this equivalent in British thermal units per hour (Btu/h)?

    a. 234 Btu/h

    b. 2730 Btu/h

    c. 800 Btu/h

    d. More information is needed to answer this.

9. Imagine, once again, that you have built a one-room cabin, and you heat it with a wood stove. When the stove operates, the walls and furniture become warm, and they in turn transfer heat energy directly to air that comes into contact with them. This is an example of how heat can be distributed throughout a room by means of

   a. radiation and conduction.

   b. radiation and absorption.

   c. convection and conduction.

   d. convection and absorption.

10. Suppose an electric heater operates at about 1500 W, and all the electricity is converted to usable heat. How much heat energy does it produce in 15 minutes?

   a. Approximately 110 Btu

   b. Approximately 1280 Btu

   c. Approximately 5120 Btu

   d. More information is needed to answer this.

CHAPTER 2

Heating with Oil and Gas

The most common modes of centralized heating systems involve the transfer of thermal energy by circulating air, water, or steam. The most popular fuels for central heating are flammable oils and gases. These compounds are largely extracted from beneath the earth's surface, and are sometimes known as *fossil fuels*. However, to some extent they can be made from biological sources. Before we examine the three major types of fossil fuel used for home heating, let's consider the most common ways in which heat is distributed with these systems.

Forced-Air Heating

Forced-air heating can be employed in systems with any fuel source. In this type of system, air is heated by a furnace and is circulated throughout the house by a network of intakes, ducts, and vents.

HOW IT WORKS

Figure 2-1 is a functional diagram of a forced-air heating system. In this example, the air intake is inside the building. Some systems take air in from the outside instead. A few systems have dual air intakes, one inside and one outside.

Air from the *intake vent* is drawn into the furnace. The *firebox* heats this air. A *furnace blower*, also called a *fan*, pushes the heated air into a network of *ducts*, where it travels to *vents* (also called *registers*) in the rooms. Small rooms usually have one vent, but large rooms may have two or more. The vents are located at the base of a wall, or in the floor near a wall. As long as the vents are not obstructed, they distribute warm air efficiently throughout the room by convection.

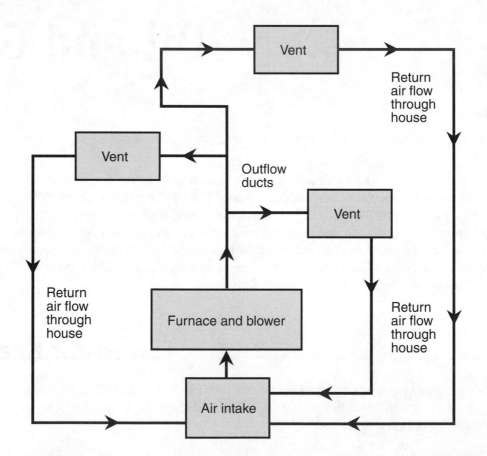

Figure 2-1 Functional diagram of a forced-air gas- or oil-fired central heating system using ducts and vents. Arrows indicate the direction of air flow.

The air intake vent, if located indoors, draws air back to the furnace from inside the house, and also to some extent from the outside, because no house is (nor should it be) perfectly airtight. If the intake vent is located outside, none of the warmed air is recirculated; positive pressure occurs inside the house, and some warm air is lost to the outside through air leaks. An indoor intake vent offers better heating efficiency than an outdoor intake vent, because the indoor air is already prewarmed by the previous passage through the furnace. An outdoor intake vent provides more fresh air and also reduces the risk of carbon monoxide (CO) gas buildup in the house.

The *exhaust* from the furnace, which contains noxious gases, is vented outside, either through a chimney or through a side vent (not shown in Figure 2-1). Extreme care should be exercised to ensure that the exhaust vent is never obstructed. In the winter, it should be frequently checked to ensure that it is not blocked by snow or ice. Obstruction of the exhaust vent can result in CO gas accumulation indoors, even if fresh air enters by means of an outdoor air intake vent. Modern high-efficiency furnaces are designed to shut down if the exhaust vent becomes partially or completely blocked.

ADVANTAGES OF FORCED-AIR HEATING

- If you leave the house for a few days, you can turn the thermostat down to a low setting (just warm enough so the water pipes won't freeze), and when you return, set it back up again, and the house will rapidly rewarm.

- You can set the thermostat low at night when you are sleeping, and set it higher during the daytime, and again, the house will rewarm quickly after the setting is raised.

- A forced-air heating system can be operated in conjunction with a humidifier in dry climates, minimizing problems with electrostatic charge buildup ("static electricity"), and maintaining a healthy indoor environment.

- In damp climates, a forced-air heating system, used without a humidifier, tends to dry the air, discouraging condensation (particularly in cool basements) and the growth of mold.

- In a forced-air system that uses an indoor air intake vent, a cleaning filter or air ionizer can be employed to remove particulates and allergens from the air.

- Forced-air furnace ductwork can serve as air conditioning (cooling) ductwork during the warmer months.

- If you have a forced-air heating system, its fan can be used in conjunction with a wood, corn, or coal stove to heat your whole house if there is an interruption in the normal fuel supply. Such an alternative system, if located near the blower intake, can also "help out the main furnace" during extremely cold weather.

LIMITATIONS OF FORCED-AIR HEATING

- A forced-air system that draws air in from a dusty outdoor environment will introduce dust into the home. Air filters can get rid of some, but not all, of this dust. This problem can be mitigated by using an indoor air intake vent rather than an outdoor one. It also helps to set the fan to "automatic" mode so it runs only when the furnace burners are aglow.

- The air filters in the furnace must be replaced frequently, or the entire system will become inefficient because it will have to work harder to circulate the air.

- If the fan fails, you will have no heat. It's important to have a 24-hour-service contract with a vendor whom you know will always be available and have parts for your particular furnace on hand. (That is a good idea, of course, with any home heating system.)

- In the event of any malfunction that causes CO gas to enter the indoor circulation, that gas will be rapidly distributed throughout the house.

PROBLEM 2-1

Why should a house not be completely airtight? Wouldn't a forced-air system that draws air from the inside, in a completely airtight house, be energy efficient, and therefore a good thing?

SOLUTION 2-1

It is true that an airtight house is more energy efficient, all other factors being equal, than a house that has a lot of air leaks. But problems can occur in houses that are too airtight. If there is a furnace malfunction that introduces CO into the circulated air, that gas may reach deadly levels before you can react, even if you have sensors with alarms installed. The danger is especially great if the event occurs while you are asleep. Besides that, open flames (with a gas cooking stove, for example) can significantly reduce the oxygen content of the air in an airtight house.

Boilers, Radiators, and Subflooring

Updated versions of "old fashioned" hot-water and steam heating systems are becoming increasingly popular. This is especially true of a new technology known as *embedded radiant heating*. A popular variant of this is known as *radiant heat subflooring*.

HOW THEY WORK

In theory, any technology that can get water to boiling or near-boiling can operate a hot-water or steam heating system. Methane, oil, and propane heaters are ideal for this purpose. Some older buildings used coal-burning systems to fire their boilers before they were converted to oil. Alternative fuels such as wood or corn can also be used to operate water heaters or boilers.

Figure 2-2 is a functional diagram of a system in which hot water or steam is piped throughout the house. In the rooms, the hot water or steam passes through

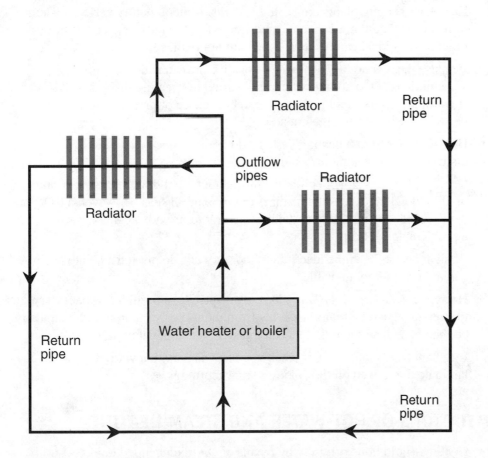

Figure 2-2 Functional diagram of a hot-water or steam heating system using pipes and radiators. Arrows indicate the direction of water or steam flow.

complex metal structures designed for minimum internal volume and maximum exposed surface area. This optimizes the transfer of heat energy from the water or steam to the surrounding environment. If the structures are directly exposed to the air, they are called *radiators*. If they are embedded in the floor and/or walls, they are called *coils*.

In a steam system, after heat energy has been lost to the air by means of the radiators, the vapor condenses and returns to the boiler as hot water. The boiler reboils the water and sends it on its way through the house again as steam. In a hot-water system, the water returns to the water heater at a lower temperature than that at which it left, and emerges from the heater ready for another round.

ADVANTAGES OF HOT-WATER AND STEAM HEATING

- Hot-water heating plants are *closed systems*. A theoretically perfect system of this kind would not consume any water after the initial charging. In practice, this ideal can be approached but not realized.

- Because there is no fan, positive or negative pressures do not build up. This minimizes the amount of energy wasted in "heating the out-of-doors" if warm air escapes and/or cold air enters because of pressure differences between the inside and the outside.

- Hot-water and steam heating systems do not introduce dust into, or distribute dust throughout, a building.

- Deadly CO gas is slow to circulate throughout a building in the event of a malfunction that causes the heating unit to emit this gas. However, a CO detector must be placed near the heating unit to provide advance warning if a problem does occur.

- Radiant heat coils embedded in the floor or walls do not intrude into rooms, and are completely invisible.

- Hot-water *baseboard radiators* have a low profile, although it is necessary to keep combustible materials away from them. Some clearance should also be maintained, so the radiators can transfer heat adequately into rooms.

- If radiant heat subflooring is used, you can wake up to a warm hardwood or laminate floor, even on the coldest winter mornings.

LIMITATIONS OF HOT-WATER AND STEAM HEATING

- Heating a cold house entirely by means of warm embedded objects is a slow process.

- Hot water can leak if radiators, coils, or pipes rupture, rust, or fracture. This can cause water damage to surrounding objects and structures.

- Steam radiators will also leak if the pipes rupture, rust, or fracture. This can damage nearby objects, and can cause serious burns to people.

- Older steam radiator systems are notorious for hissing and clanging as the pipes expand and contract. Noises of this sort are rare in well-designed, properly operating, newer hot-water systems.

- Any type of radiator becomes hot to the touch and can burn a person who comes into direct contact with it.

- In a hot-water system, the water must be kept free of minerals to prevent deposits from building up in the pipes, water heater, and/or boiler. This requires the use of a good water softener (or in the ideal case, distilled water).

PROBLEM 2-2

Can radiant heat subflooring be used if the floors are carpeted?

SOLUTION 2-2

Yes, provided that the carpet is installed with matting underneath that does not provide too much thermal insulation. Usually, this requires the expertise of a professional installer.

Oilheat Technology

Many homes rely on oil for heating, particularly in the northeastern United States. *Oilheat technology* has undergone a renaissance in recent years. Engineers have developed high-efficiency, clean-burning oilheat systems that compete favorably with systems that use other types of fuel.

HOW IT WORKS

An oil-fired central heating system consists of several components. A *fuel tank* is located on the property. This tank can be either above ground or below ground, depending on the location, the ordinances and covenants in the neighborhood, and the preference of the property owner. A *pipeline* runs from the tank to the furnace unit. The tank is periodically filled by service personnel.

In the furnace, heating oil is *atomized*, in a manner similar to the way gasoline is atomized in the carburetor of a motor vehicle. The process consists of breaking the oil into fine particles that are mixed with air. The oil/air mixture is set on fire by an *electric igniter*. The resulting flames release heat in a *combustion chamber*. This heat can be used with a forced-air distribution system, a water heater and radiator system, or a steam boiler and radiator system.

In older homes, steam radiator systems are often found with oilheat furnaces. In newer homes with oilheat technology, radiant heat subflooring, in conjunction with a water heater, is popular. Some homes employ hot-water baseboard radiators consisting of pipes with steel plates attached at right angles, maximizing surface area while minimizing volume (see Figure 2-3). These radiators are run along the floor at the base of a wall. They are covered by metal registers to protect them against dust accumulation, and to prevent physical damage to the plates if the assembly is bumped. The registers also reduce the risk of burns to people who accidentally come into contact with them.

The main byproducts of oil combustion are water vapor (H_2O) and carbon dioxide (CO_2). Trace amounts of sulfur dioxide (SO_2), CO, and particulate matter are also produced. All of these waste materials are vented to the outside as exhaust. This can be done through a conventional chimney flue, or, with high-efficiency oilheat systems, through a plastic *flue pipe* passing through an outer wall of the house.

ADVANTAGES OF OIL HEATING

- In an oilheat system, there is no pipeline to a central supplier. The tank can be located on the property. This can be significant for people who want to live in rural or remote areas, far from urban centers.

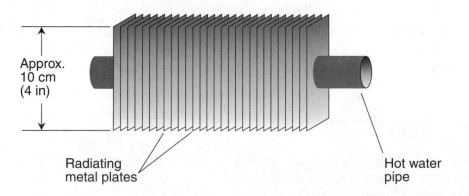

Figure 2-3 Simplified diagram of a section of a hot-water baseboard radiator. The elements are often covered by vents to protect them, and to protect people and objects near them.

- Oil is a relatively safe fuel. An oil leak can produce a mess, but it does not pose the danger of an explosion. Oil is less explosive than any flammable gas, and is also less explosive than gasoline, alcohol, or kerosene.

- If an oil-fired system malfunctions, smoke is almost always generated. This calls attention to the potential for a problem in an immediate and dramatic way.

- Oil is a high-density fuel. That means a small volume and mass of oil contains a large amount of *potential energy*. An onsite oil tank can be modest in size, and yet provide several weeks' worth of central heat for a household.

- Because of its high density, oil is portable. Small oilheat systems are practical for use in recreational vehicles, hunting cabins, and other small, temporary living spaces.

- Heating oil can be mixed with *biofuels* such as surplus vegetable oil, used cooking grease, or discarded animal fat. Oilheat systems can be designed to efficiently burn such hybrid fuels.

LIMITATIONS OF OIL HEATING

- The onsite storage tank must be carefully watched so that it will not go empty at any time during the cold season. If it does, you had better have a backup heating system ready to take over until the oil supplier can come out!

- The price of heating oil is directly related to the price of crude oil. This price can spike rapidly. The price prognosis for crude can be debated, but we should not be surprised if it continues to rise over the long term.

- Much of the world's crude oil comes from countries with a history of political instability. As a result, there is an ever-present risk of a sudden reduction in the supply.

- Temporary reductions in the crude oil supply (for the United States, at least) can be caused by massive hurricanes in the Gulf of Mexico.

- The world's supply of crude oil is finite, and it is not renewable. (Biofuels, mentioned above, are renewable.)

- Certain components in modern oilheat systems require electricity. If you want the furnace to work during an electrical power failure, a backup generator must be available, ready to go at any time, capable of delivering a clean sine wave (so the electronic control circuits will work), and able to provide enough current to operate the whole furnace, including all its components. A professional should be consulted concerning the installation of a backup generator for use with any heating system.

PROBLEM 2-3
Won't oil freeze or become sluggish in cold weather, preventing free flow from the tank to the furnace?

SOLUTION 2-3
Heating oil won't freeze at temperatures encountered in any inhabited region on earth. Heating oil can become sluggish if it gets cold, just as motor oil does, but this is rarely a problem. It can be dealt with in the same way as water pipes are protected from freezing. It helps if a residential oil tank can be placed below ground, where the temperature does not vary greatly.

Methane (Natural Gas) Heating

During the 20th century, *methane* (CH_4), often called *natural gas* or simply *gas*, became the most popular fuel for central heating in the United States. (Actually, methane is only one of the constituents of true natural gas.) There were excellent technological and economic reasons for this. Natural gas occurs in abundance beneath the surface of the earth, and it is easily tapped. During the early 21st century, periodic shortages began to occur. Methane nevertheless remains a viable heating fuel choice, particularly in urban centers where gas pipeline systems are in place.

HOW IT WORKS

The geometric configurations of gas-burning furnaces vary greatly, but the basic principles of operation are similar. Here is a simplified description of the operation of a conventional forced-air gas furnace.

The *thermostat* determines when heat is called for. It is a simple switch that starts the process of furnace operation, and is located in the main living space of the house. The *induced draft motor*, also known as a *combustion air blower*, supplies the air necessary to burn the fuel, and also blows away the exhaust gases resulting from combustion. These gases can be vented out through a chimney flue. In a high-efficiency furnace, these gases are *condensed* to extract additional heat, and the cooled exhaust is vented out of the house through a plastic flue pipe.

The combustion air blower starts running as soon as the thermostat calls for heat. A *centrifugal switch* verifies that the combustion air blower is operating. After a short delay, the *gas valve* opens, and an electric igniter, also known as a *glow plug*, sets the fuel on fire. In older systems, a *pilot light* is used instead of an electric igniter, and a *thermocouple*, which is a temperature-actuated electronic switch, keeps the gas valve from opening unless the pilot light is burning. The *burners* provide the combustion that heats the firebox or *heat exchanger*.

In a forced-air system, the furnace fan (furnace blower) circulates the warmed air from the firebox or heat exchanger throughout the house. It starts after a delay period of 1 to 3 minutes following the ignition of the burners. When the temperature in the house rises to a certain level, the gas valve closes, and the burners go off. Then the combustion air blower stops running. After a delay of 1 to 3 minutes, the fan shuts down if the system is set for *automatic (auto) fan mode*. In most forced-air gas heating systems, the operator has the option of leaving the fan on constantly. This can minimize thermal stratification, where cold air tends to accumulate in certain rooms, or near the floors of all rooms. If a wood-, corn-, or coal-burning stove is located near the gas furnace air intake, and if the fan is left running continuously, the stove can be used to heat the house as an alternative, or as a supplement, to the furnace.

Some gas furnaces are used with water heaters or boilers. These systems often employ multiple thermostats to regulate the temperature in individual rooms or portions of the house. The thermostats actuate valves that control the flow of water to the various radiators or coils.

ADVANTAGES OF METHANE HEATING

- Methane heating systems are efficient. Modern furnaces with exhaust condensers can convert nearly all of the potential energy in the methane into usable heat.

- High-efficiency gas furnaces can vent exhaust directly through a wall to the outside. This eliminates the need for a chimney. When a high-efficiency gas furnace is installed in place of an oilheat system or an older gas furnace, the chimney flue for the displaced system can be used with a supplemental heat source such as a wood-, corn-, or coal-burning stove, as long as that flue is not used to vent any other heating appliance or system.

- Natural gas is readily available in most cities and towns. An uninterrupted supply is provided by means of underground pipelines. There is no need to worry about the status of an onsite supply tank.

- Gas furnaces (as well as gas stoves, gas fireplaces, and gas water heaters) burn clean. They produce relatively little air pollution, and essentially no smoke.

- Forced-air gas furnaces are available in *upflow*, *downflow*, and *horizontal* (or *lateral*) *flow* designs. This allows them to be installed in basements, modular or mobile homes, and attics or crawl spaces, respectively.

LIMITATIONS OF METHANE HEATING

- Methane leaks can cause flash fires or explosions. (This is true of any fuel burned in a gaseous state, including hydrogen.) In most locales, methane gas is given an artificial scent that is easy to recognize. This can alert people to the existence of gas leaks. If you "smell gas" in your house, *get out of there immediately* and call the fire department.

- In recent years, the price volatility, and occasional problems with the supply, of natural gas has marred its reputation as the most reliable home heating fuel.

- In rural locations, and in less developed parts of the world, gas pipelines are not in place.

- Methane gas is not easy to store for a single household, unless you own a large farm or ranch and the local laws or covenants allow methane storage tanks on private property.

- The world's supply of naturally occurring methane is finite, and it is not renewable. (However, methane can be derived from certain biological processes, and this source is renewable.)

- Certain components in modern methane heating systems require electricity. They will not work if the electrical power fails, unless a generator is available. See the last limitation note for oilheat systems, above.

PROBLEM 2-4
Can other gases be used in place of methane in a heating system designed to burn methane? How about hydrogen, in particular?

SOLUTION 2-4
Hydrogen holds some promise for eventual use in place of methane as the piped-in fuel in cities and towns. But hydrogen, being the lightest chemical element, leaks more easily than methane. In addition, it burns hotter. This will necessitate changes in pipeline and furnace design. As of this writing, even though hydrogen is the most abundant element in the universe, it does not occur in its free form on earth in quantities that make it easy to utilize as fuel. Numerous methods exist to produce hydrogen by separating it out from chemical compounds containing it, but at the time of this writing, all of these technologies are expensive and inefficient.

Propane Heating

Propane is the term for *liquefied petroleum* (LP) gas. In the United States, this consists mainly of the hydrocarbon propane (C_3H_8), hence the name. In some countries, a mixture of propane and *butane* (C_4H_{10}) is used. Propane, in its gaseous state, is heavier than methane, and butane is heavier than propane.

HOW IT WORKS

Propane can be obtained as a byproduct in the process of extracting methane from natural gas, or as a byproduct in the process of refining crude oil. Propane can be stored under pressure in liquid form. When the liquid is released from the pressurized tank into the atmosphere, it becomes a flammable gas similar to methane. Propane is colorless and odorless, as are butane and methane, in the gaseous state. A compound called *ethyl mercaptan*, which smells like rotten eggs, is added to propane so leaks can be easily detected.

The fact that propane can be stored in tanks makes it portable. For this reason, it is a popular alternative fuel for home backup generators, generators for use with recreational vehicles, and portable cooking stoves. It can be used in a central home heating system as an alternative to oil or methane. Propane has also been used as a fuel for motor vehicles. When propane is depressurized by releasing it from a tank, it burns in much the same way as methane.

ADVANTAGES OF PROPANE HEATING

- Propane can be used almost anywhere, because it can be stored in tanks. When you see a silver tank shaped like a fat sausage in the yard of a rural home, farm house, or ranch house, you can be reasonably sure that it's a propane tank.

- Propane is not soluble in water. For this reason, it presents little risk of contamination to ground water or soil.

- Propane heating systems are less polluting than wood- or coal-burning systems, and are comparable to oil and methane systems in this respect.

- As an alternative fuel, propane may offer cost savings to people who live in areas where methane and oil prices are high.

- With a propane system, all the energy needs of a home can be met, and a home can be operated off the utility grid. Propane can supply the energy for the furnace, the electrical system (using a propane generator), a water heater, a range and oven, a refrigerator, and even a laundry washer and dryer.

- Propane is not as flammable as methane, so it poses less of an explosion hazard. However, the same precautions should be taken with propane-burning systems as are necessary with methane-burning systems.

LIMITATIONS OF PROPANE HEATING

- The price of propane may not be the lowest price for any fuel available in a given location. When considering price, all factors must be considered, including the cost of furnace conversion or installation (if required) and maintenance, the risk of interruptions in the fuel supply, and the attitudes of homeowner's insurance companies (reflected by their rates) toward the use of various fuels in a particular location.

- Propane, while stored in liquefied form, does not have the highest energy content per volume of any conventional fuel choice. That distinction goes to home heating oil. Because of this fact, propane storage tanks must be larger, per unit of energy delivered, than oil storage tanks.

- If the tank temperature falls to extremely low levels (approximately −34°C or −29°F), propane may not revert to the gaseous phase when released. This will cause the heating apparatus to become inoperative.

- Not everyone can detect the odor of ethyl mercaptan in LP gas.

- If tanks are allowed to run completely empty, or if oxidation (rust) occurs inside the tank or pipeline, *odor fade* may occur. This means that you might not be able to detect a propane leak based on your sense of smell alone. *Propane gas detectors* are recommended in any propane installation for this reason.

- Certain components in modern propane heating systems require electricity. They will not work if the electrical power fails, unless a generator is available. See the last limitation note for oilheat systems, above.

PROBLEM 2-5
All of the fuels discussed so far involve "burning stuff in the house." When are we going to get away from that? Combustion of any flammable material, except for hydrogen, pollutes the air. Some of these fuels are obtained by literally poking holes in the earth. Why can't we free ourselves from this ancient business, and start thinking about the long-term future?

SOLUTION 2-5

In order to be objective and complete, this book discusses existing, as well as evolving and futuristic, methods of harnessing energy for human needs. Most scientists, economists, and even industry executives agree that the old ways must and will pass. But, wishful thinking aside, "archaic" energy technologies will remain the most practical alternatives for much of the world's population for some time to come.

Quiz

This is an "open book" quiz. You may refer to the text in this chapter. A good score is eight correct. Answers are in the back of the book.

1. If you want to operate a backup generator to run the fan for a forced-air propane furnace when the electrical power fails, which of the following must be done?

    a. Methane must be used instead of propane to fuel the furnace.

    b. Butane must be used instead of propane to fuel the furnace.

    c. The backup generator must be able to provide power to the fan, any existing control circuits, and an igniter (if the system has one), all at the same time.

    d. Forget it! A propane forced-air heating system must never be powered by a backup generator.

2. Which of the following fuels is the least explosive?

    a. Heating oil

    b. Propane gas

    c. Methane gas

    d. Hydrogen gas

3. Which of the following is an advantage of a forced-air heating system?

    a. The fan can be used with a wood-burning stove to distribute its heat throughout the house.

    b. A cool house can be warmed up rapidly.

    c. The ductwork can be used for air conditioning in the summer.

    d. All of the above

4. How can the efficiency of a methane furnace be improved or optimized?

   a. The exhaust can be condensed, extracting additional heat from it.

   b. Atomized oil can be burned in the firebox instead.

   c. Atomized gasoline can be burned in the firebox instead.

   d. The onsite methane supply tank can be placed below ground.

5. One of the most significant advantages of a modern hot-water heating system that employs radiant heat subflooring is the fact that

   a. it can heat up a cold house fast.

   b. it cannot produce CO gas.

   c. you get to wake up to warm floors in the winter.

   d. the only exhaust byproduct is liquid water.

6. A pipeline to a central supplier is usually required in

   a. a propane heating system.

   b. a methane heating system.

   c. an oilheat system.

   d. All of the above

7. In an older gas furnace with a pilot light, what type of electronic device detects heat from the pilot flame to ensure that the gas valve cannot open if the pilot light has gone out?

   a. A heat exchanger

   b. An igniter

   c. A thermocouple

   d. Nothing. You'd better check that pilot light often!

8. With a forced-air heating system, an outdoor air intake vent

   a. offers the advantage of providing a constant supply of fresh air to the interior as long as the fan is blowing.

   b. recirculates the indoor air, thus reducing the efficiency of the system and running the risk of CO buildup.

   c. is an ideal arrangement because it makes CO buildup impossible, no matter what else may happen in the system.

   d. should never be used, because it defeats the purpose of the system by drawing cold air into the house.

9. Hydrogen gas may eventually be used in place of methane for gas heating systems, provided that

    a. the helium-to-hydrogen conversion process can be perfected.

    b. hydrogen can be obtained in abundance at reasonable cost.

    c. a way can be found to burn hydrogen without producing CO gas.

    d. all of the above things happen.

10. Which of the following heat distribution methods, if any, is *incompatible* with oilheat technology?

    a. Radiant heat subflooring

    b. Hot-water radiators

    c. Forced air

    d. All of the above methods are compatible with oilheat technology.

CHAPTER 3

Heating and Cooling with Electricity

Electricity can be used for heating and cooling the indoor environment, either directly or indirectly. Before we get into how electric heating and cooling systems work, let's review the ways scientists define how hot or cold things are.

Temperature

Temperature is a quantitative expression of the amount of *kinetic energy* (or heat energy) contained in matter. When energy is allowed to freely move from one medium into another in the form of heat, the temperatures of the two media tend to equalize. This process is known as *heat entropy*. An electric heating or cooling system acts against this natural process on a localized scale.

THE CELSIUS (OR CENTIGRADE) SCALE

If you have a sample of ice and you place it where the temperature is above the freezing point, the ice starts to melt as it accepts heat from the environment. The ice, and the liquid water produced as it melts, is at a temperature value of 0°C. As energy flows into the chunk of ice, more and more of it melts. If energy continues to flow into the water once it has become all liquid, its temperature begins to increase. Eventually, a point is reached where the water starts to boil, and some of it changes to the gaseous state. The liquid water temperature, and the water vapor that comes immediately off of it, is at 100°C. As energy flows into the water, more and more of it evaporates. If energy continues to flow into the water once it has all become vapor, its temperature begins to increase again. Ultimately, the only limit to how hot the water vapor can become depends on how much energy the heating elements can deliver into it.

There are two definitive points for water—the *freezing point* and the *boiling point*—at which there exist two specific numbers for temperature. We can define a scheme to express temperature based on these two points. This scheme is the *Celsius temperature scale*, named after the scientist who supposedly first came up with the idea. Sometimes it is called the *centigrade temperature scale*, because one degree (1° or 1 deg) of temperature in this scale is equal to 1/100 of the difference between the melting temperature of pure water at sea level and the boiling temperature of pure water at sea level. The prefix multiplier "centi-" means 1/100, so "centigrade" literally means "graduations by the hundredth part."

THE KELVIN SCALE

It is possible to freeze water and keep cooling it down, or boil it all away into vapor and then keep heating it up. Temperatures can plunge far below 0°C, and can rise far above 100°C.

There is an absolute limit to how low the temperature in degrees Celsius can become, but there is no limit on the upper end of the scale. We might take extraordinay efforts to cool a chunk of ice down to see how cold we can make it, but we can never chill it down to a temperature any lower than −273.15°C. This is known as *absolute zero*. An object at absolute zero cannot transfer energy to anything else, because it possesses no energy to transfer. There is believed to be no such object in our universe, although some atoms in the vast reaches of intergalactic space come close.

Absolute zero is the basis for the *Kelvin temperature scale*. Units in this scale are known as *kelvins* (K). A temperature of −273.15°C is equal to 0 K. The size of the kelvin increment is the same as the size of the Celsius increment. Therefore, 0°C = 273.15 K, and +100°C = 373.15 K.

On the high end, it is possible to keep heating matter up indefinitely. Temperatures in the cores of stars rise into the millions of kelvins. In the centers of galaxies, quasars, and other extreme celestial objects, perhaps temperatures reach billions (thousand-millions) of kelvins.

THE FAHRENHEIT SCALE

In much of the English-speaking world, and especially in the United States, the *Fahrenheit temperature scale* (°F) is used by lay people. The Fahrenheit increment is precisely 5/9 as large as the Celsius increment. The melting temperature of pure water ice at sea level is +32°F, and the boiling point of pure liquid water is +212°F. Thus, +32°F corresponds to 0°C, and +212°F corresponds to +100°C. Absolute zero on the Fahrenheit scale is approximately −459.67°F.

Let F be the temperature in °F, and let C be the temperature in °C. In order to convert from degrees Fahrenheit to degrees Celsius, use this formula:

$$C = (5/9)(F - 32)$$

To convert a reading from degrees Celsius to degrees Fahrenheit, use this formula:

$$F = 1.8C + 32$$

PROBLEM 3-1
What is the Celsius equivalent of a temperature of 72°F? Round the answer off to the nearest degree.

SOLUTION 3-1
To solve this, simply use the above formula for converting Fahrenheit temperatures to Celsius temperatures:

$$C = (5/9)(F - 32)$$

So in this case:

$$C = (5/9)(72 - 32)$$

$$= 5/9 \times 40 = 22°C$$

PROBLEM 3-2
What is the Kelvin equivalent of a temperature of 80°F? Round the answer off to the nearest whole number.

SOLUTION 3-2

First, convert from degrees Fahrenheit to degrees Celsius:

$$C = (5/9)(80 - 32)$$

$$= 5/9 \times 48 = 26.67°C$$

Let's not round our answer off yet, because we have another calculation to perform. Remember that the difference between readings in the Celsius and Kelvin scales is always equal to 273.15. The Kelvin figure is the greater of the two. So we must add 273.15 to our Celsius reading. If K represents the temperature in kelvins, then:

$$K = C + 273.15$$

$$= 26.67 + 273.15$$

$$= 299.82 \text{ K}$$

We can round this off to 300 K.

STANDARD TEMPERATURE AND PRESSURE (STP)

To set a reference for temperature and pressure, against which measurements can be made and experiments conducted, scientists have defined *standard temperature and pressure* (STP). The standard temperature is 0°C (32°F), the freezing point or melting point of pure liquid water. Standard pressure is the atmospheric pressure that will support a column of mercury 0.760 m (a little less than 30 in) high at sea level. This is 14.7 pounds per inch squared (lb/in²), which translates to approximately 1.01×10^5 newtons per meter squared (N/m²).

We don't think of air as having significant mass, but that is because we're immersed in it. The density of dry air at STP is approximately 1.29 kg/m³. A parcel of air measuring 4.00 m high by 4.00 m deep by 4.00 m wide, the size of a large bedroom with a high ceiling, masses 82.6 kg. In the earth's gravitational field, that's just about the weight of a full-grown man.

Electric Resistance Heating

The most straightforward way to obtain heat from electricity is to apply a *voltage* to a resistive element, causing *current* to flow through it. When this is done, heat is generated in the form of *infrared* (IR) radiation. In some cases, visible and/or *ultraviolet* (UV) radiation is produced as well. *Electric resistance heating* is sometimes used in regions where the winters do not get particularly cold. It does not work well as a stand-alone heating system where winters are severe, although it can serve as a supplemental heat source to aid the main furnace. Some seasonal cabins use electric resistance heating to take the chill out of the indoor air on spring and fall nights.

HEAT, VOLTAGE, AND RESISTANCE

Figure 3-1 is a functional diagram of an electric resistance heating unit. The alternating-current (AC) utility electricity provides the power. The *heating element* is the heart of the system, usually consisting of a large coil or set of coils made of resistive wire. This wire is capable of withstanding high temperatures without melting or breaking.

The amount of heat power produced by a resistive element depends on the voltage supplied to it, and also on its resistance. Voltage in household AC circuits is expressed in terms of the *effective voltage*, more often called the *root mean square* (rms) *voltage*. This is the AC voltage that produces the same amount of heat, if applied to a resistive element, as a *direct current* (DC) voltage of the same numerical

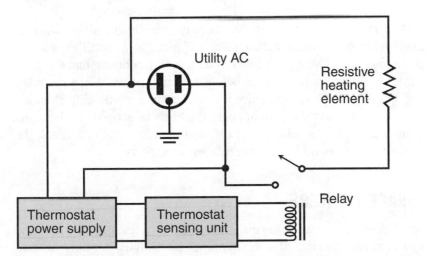

Figure 3-1 Functional diagram of a thermostat-controlled electric resistance heating system designed for use in a single room.

value would produce. If E is the rms AC utility voltage in *volts* (V), and if R is the resistance of the heating element in *ohms* (Ω), then the power P_W dissipated by the element, in *watts* (W), is given by this formula:

$$P_W = E^2/R$$

The typical rms household voltage in the United States is either 117 V AC or 234 V AC, plus or minus a few percent.

HEAT, CURRENT, AND RESISTANCE

The amount of heat power produced by a resistive element can also be determined if the current I that the element draws from the utility in *amperes* (A) is known along with the resistance in ohms. In this case:

$$P_W = I^2R$$

This formula, and the preceding one involving voltage, is valid if and only if all the power supplied to the resistive element is converted to heat. In common heaters, that is always the case, even if the element glows red-hot.

The above formulas can be modified to express heat power in British thermal units per hour (Btu/h). In either case, multiply the power in watts by 3.41 to obtain the "Btu per hour power," $P_{Btu/h}$:

$$P_{Btu/h} = 3.41\ E^2/R$$

$$P_{Btu/h} = 3.41\ I^2R$$

In Figure 3-1, a *thermostat* and its power supply are shown, along with an electromechanical *relay*. The thermostat contains a *bi-metal strip* that flexes as the temperature rises and falls. This opens and closes a set of contacts that supplies current to the relay if the temperature falls below a certain point. When the relay receives current, its contacts close, and current flows through the heating element. When the temperature rises above a certain point, the bi-metal strip flexes back, its contacts open, the current to the relay is cut off, and the relay contacts break the circuit so the resistive element no longer receives any electricity.

ELECTRIC SPACE HEATERS

For heating a small room, or for warming up a large room by a few degrees, portable *electric space heaters* are available. There are numerous designs. All have heating elements with on/off power switches. Some have fans that draw cool air in from the back or the bottom and blow heated air out the front (see Figure 3-2, part A). Others

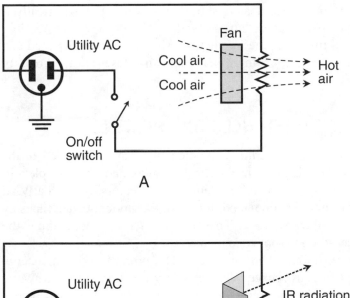

A

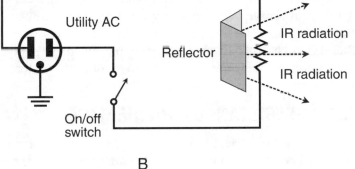

B

Figure 3-2   Simplified diagrams of portable electric space heaters. In diagram A, a fan blows heated air into the room. In B, a reflector directs IR radiation into the room. (The switches are shown in the open position, but of course the heaters operate only when the switches are closed!)

take advantage of IR radiation from large elements with reflectors behind them (see Figure 3-2, part B).

Not all electric space heaters are well designed. Some are not physically large enough to radiate away all the heat power that they generate. As a result, they can overheat and may actually damage themselves. The best electric space heaters are rather massive, bulky, and have three-wire, heavy-duty cords. Most have exteriors made entirely of metal. Some have plastic casings with metal reflectors inside and a metal grate or grille to prevent users from coming into contact with the elements. Some units have elements wound into, or around, ceramic forms. These days, most (but not all!) portable electric space heaters shut themselves off automatically if they overheat or are tipped over. Let the buyer beware!

When using a heater of this kind, the instructions should be carefully read and heeded in every detail. Neglect or improper use can cause burns, fires, or electrocution.

RADIANT ELECTRIC RESISTANCE ZONE HEATING

In *radiant electric resistance zone heating*, the elements can be installed in the ceiling. This ensures that there is little chance for obstruction, and that people will not accidentally come into contact with hot elements. Reflectors direct nearly all the heat energy downward into the room, and minimize the amount that escapes by conduction through the ceiling or roof. Thermostats in each room allow the temperature to be set according to room occupancy.

This system works even though "hot air rises." The primary mode of energy transfer is IR radiation, which can occur equally well in any direction, even downward. The IR energy from the ceiling elements warms the people and objects below. Ultimately, all of the IR energy is absorbed by objects and people, and they heat the air by convection and conduction, but these are secondary modes of heat transfer.

BASEBOARD ELECTRIC RESISTANCE ZONE HEATING

Wall baseboard radiators can be used for electric zone heating. These radiators are similar to those employed in hot-water or steam systems. The same principles apply to electric systems of this kind as apply to those using other sources of fuel. The main advantage of electric heating in this application is that it does not pose any danger of structural damage as a result of water or steam leakage.

Electric baseboard radiators are usually installed on exterior walls at floor level underneath windows. Heat is transferred to the air in the room mainly by convection. There is a small gap between the bottom of the radiator and the floor, allowing cool air to be drawn in so it can pass through the radiating fins and then rise upward along the wall. The fins are enclosed in metal registers to protect them from damage, and to reduce the chance that people will be burned, or objects will catch fire, by contact with them. Nevertheless, plenty of clearance should be allowed between baseboard radiators and objects such as furniture and draperies.

CENTRAL ELECTRIC RESISTANCE HEATING

Some furnaces use electric resistance heating. Instead of a firebox where combustion occurs, high-wattage resistance elements are installed.

An electric furnace usually operates at 234 V AC. For a given resistance in the heating elements, this translates into four times the power as that from a 117 V AC system. Recall the formulas for heat power in terms of voltage and resistance:

$$P_W = E^2/R$$

$$P_{Btu/h} = 3.41\ E^2/R$$

From these formulas we can see that if the voltage (E) is increased by a factor of 2, then the power (P_W or $P_{Btu/h}$) is increased by a factor of 2^2, or 4. This is only valid, of course, if the resistive heating elements can withstand the doubling of current that occurs, and can radiate away the fourfold increase in dissipated power.

Electric furnace heating elements are rated at several kilowatts each, and two or more of them are connected together in parallel to obtain the heat for the entire house or building. Forced air is the mode of choice. Figure 2-1 (from the previous chapter) can serve as a simplified functional diagram of a forced-air electric resistance heating system. The furnace itself is, in effect, an oversized electric space heater, as diagrammed in Figure 3-3.

ADVANTAGES OF ELECTRIC RESISTANCE HEATING

- No exhaust gases or waste byproducts are created in or around the house.
- There is no risk of CO poisoning if the system malfunctions.
- There is no explosion hazard, because no flammable gases are used.
- In zone heating systems, individual thermostats can be located in each room, and temperatures adjusted according to room usage.
- In radiant zone heating systems, dust is not circulated throughout the house.
- Breakers or fuses can provide a cheap, automatic shutdown mechanism in the event of a short circuit, as long as the electrical system is properly grounded.

LIMITATIONS OF ELECTRIC RESISTANCE HEATING

- The dissipation of power by forcing current through a resistance is an inefficient way to heat a room or building, all things considered. All of the electricity is converted to heat in the unit, but the electric generating plants, from which the power originates, are less efficient than a typical gas- or oil-fired furnace.

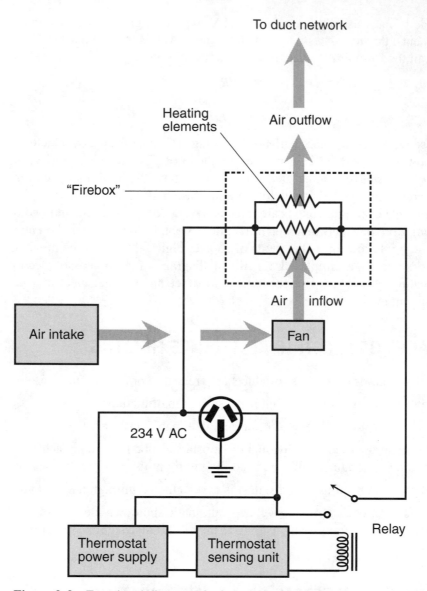

Figure 3-3   Functional diagram of a forced-air electric furnace. Arrows indicate the direction of air flow.

• Because of the overall inefficiency of electricity as a heating source, an electric system pollutes the environment more, all stages of the process considered, than a gas- or oil-fired system of equal capacity.

- If power fails, electric resistance heating won't work, and onsite generators are impractical because of the high current demand.
- Electric space heaters pose a fire hazard if not well made or if improperly used.
- Electrocution can occur if an improperly grounded electric space heater develops a short circuit.
- Electric resistance is not practical for use in a main heating system where the winters are severe, unless the electric utility rates are exceptionally low.

PROBLEM 3-3

Suppose a heating unit operates from the standard 117 V AC utility electricity common in American households. If the resistance of the heating element is 18.25 Ω when it is connected to the electric current source and is fully heated up, how much heat power does it produce in watts? How much heat power does it produce in British thermal units per hour? Round off the answers to the nearest whole unit.

SOLUTION 3-3

To determine the heat power in watts, use the formula for *wattage* in terms of AC voltage and resistance:

$$P_W = E^2/R$$

$$= 117^2/18.25$$

$$= 750 \text{ W}$$

To determine the heat power in British thermal units per hour, multiply the heat power in watts by 3.41, as follows:

$$P_{Btu/h} = 3.41 \times 750$$

$$= 2558 \text{ Btu/h}$$

Principles of Cooling

In descriptions of the way cooling systems work, the calorie (cal), defined in Chapter 1, is often used as a unit of energy. This is because pure liquid water requires, or releases, exactly one *calorie per gram* (1 cal/g) to warm up or cool down by 1°C. A calorie per gram is the equivalent of 4.184 *joules per gram* (J/g), or 4184 *joules per kilogram* (J/kg).

EVAPORATION AND CONDENSATION

The above-described characteristic of water, called the *specific heat*, is a valid description of its behavior only as long as it remains entirely liquid during the heating or cooling process. If liquid water changes to vapor (evaporates), it acquires heat energy from the surroundings, and the surroundings therefore lose heat energy, becoming cooler. Conversely, when water vapor changes to liquid (condenses), the water transfers heat to the surroundings, and the surroundings gain heat energy, becoming warmer. The same sort of thing happens with other compounds that change state from liquid to vapor or vice versa.

Suppose a kettle of water is sitting on a stove top, and the temperature of the water is exactly at the boiling point ($+100°C$), but it has not started to actively boil. As heat is continually applied, boiling (rapid evaporation) begins. The water becomes proportionately more and more vapor, and less and less liquid. But the temperature of the liquid water in the kettle stays at $+100°C$. Eventually, all the liquid has boiled away, and only water vapor is left. Imagine that we have captured all this vapor in an enclosure, and all the air has been driven out, leaving pure water vapor. Suppose that the stove burner keeps on heating the water even after all of it has boiled into vapor. The vapor can become hotter than $+100°C$. The upper limit is determined only by the size and temperature of the burner.

Consider now what happens if we take the enclosure, along with the kettle, off of the stove and put it someplace where the temperature is just a couple of degrees above freezing. The water vapor begins to grow colder. The vapor temperature eventually drops to $+100°C$, at which point it begins to condense. The temperature of this liquid is $+100°C$. This process takes place until all the vapor in the chamber has condensed. The chamber keeps growing colder. Once all the water vapor has condensed into liquid, the temperature of the liquid water begins to fall below $+100°C$.

HEAT OF VAPORIZATION

Interestingly, the temperature of water does not rise or fall continuously when heating or cooling takes place near its boiling or condensation point. Instead, the water temperature follows a curve something like that shown in Figure 3-4. At A, the air temperature is getting warmer; at B, it is getting colder. The water temperature "stalls" as it boils or condenses. Other substances exhibit this same property when they boil or condense, and certain substances are notable for the large amounts of heat they take in or release when they change state in this way. Such substances are useful in systems that transfer heat energy from one place to another. These are *refrigerants* such as R-134a, one of a class of chemicals known as *hydrofluorocarbons* (HFCs) that are used in systems that transfer heat energy from one place to another.

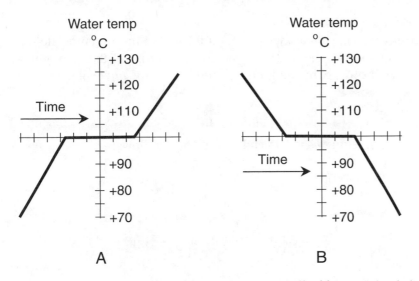

Figure 3-4   Water as it boils and condenses. At A, liquid water takes in heat from the surroundings as it vaporizes. At B, water vapor gives up heat to the surroundings as it condenses. The water temperature "stalls" as vaporization or condensation occurs.

It takes a certain amount of energy to change a sample of liquid to its gaseous state, assuming the matter can exist in either of these two states. In the case of pure water, it takes 540 cal (2260 J) to convert 1 g of liquid at +100°C to 1 g of vapor at +100°C. In the reverse scenario, if 1 g of pure water vapor at +100°C condenses completely and becomes liquid at +100°C, it gives up 540 cal (2260 J) of energy. This quantity varies for different substances, and is called the *heat of vaporization* for the substance.

If the heat of vaporization (in calories per gram or joules per gram) is symbolized h_v, the heat added or given up by a sample of matter (in calories or joules, respectively) is symbolized h, and the mass of the sample (in grams) is symbolized m, then:

$$h_v = h/m$$

PROBLEM 3-4

Suppose a certain substance changes from liquid to vapor (or vice versa) at +500°C under normal atmospheric pressure. Imagine a beaker containing 67.5 g of this substance entirely in liquid form at +500°C. If the heat of vaporization is 845 cal/g, how much heat energy, in calories, will this sample of matter absorb if it completely evaporates? Round off the answer to three significant figures, and express it in scientific (power-of-10) notation.

SOLUTION 3-4

The units are grams for m, and calories per gram for h_v. We must manipulate the above formula so it expresses the heat, h (in calories), in terms of the other given quantities. This can be done by multiplying both sides by m, giving us this formula:

$$h = h_v m$$

Plug in the numbers, letting $h_v = 845$ and $m = 67.5$, as follows:

$$h = 845 \times 67.5$$

$$= 57{,}038 \text{ cal}$$

$$= 5.70 \times 10^4 \text{ cal}$$

PROBLEM 3-5

Imagine a substance that changes from vapor to liquid (or vice versa) at +15°C under normal atmospheric pressure. Imagine an enclosed system that contains exactly 2 kg of this substance, entirely in vapor form, at +15°C. If the heat of vaporization is 8500 J/g, how much heat energy, in joules, will this sample of matter give up if it completely liquefies? Express the answer to two significant figures in scientific notation.

SOLUTION 3-5

First, we must convert the mass of the sample, 2 kg, into grams. We know that 1 kg = 1000 g, so 2 kg = 2000 g. Then we can use the same formula as in the previous problem:

$$h = h_v m$$

Plug in the numbers, letting $h_v = 8500$ and $m = 2000$, as follows:

$$h = 8500 \times 2000$$

$$17{,}000{,}000 \text{ J}$$

$$= 1.7 \times 10^7 \text{ J}$$

PROBLEM 3-6

Convert the amount of heat energy derived in the previous solution to British thermal units. Call this figure h_{Btu}. Round off the answer to the nearest 100 Btu.

SOLUTION 3-6

From Chapter 1, recall the fact that 1 Btu = 1055 J. This means we must divide the above result by 1055, as follows:

$$h_{Btu} = 1.7 \times 10^7 \, / \, 1055$$

$$= 16{,}100 \text{ Btu}$$

Electric Heat Pumps

An *electric heat pump* uses electricity to power a machine that transfers thermal energy from one place to another. The term "pump" comes from the fact that the electricity is harnessed to "push" heat energy from place to place, and not to directly generate heat energy.

AIR EXCHANGE

Figure 3-5 is a simplified functional diagram of an *air exchange heat pump*, also called an *air source heat pump*, operating to transfer heat energy from the outdoor environment to the indoor environment. This is the "heating mode." Outdoor air is drawn or blown by a fan through a coil that contains a refrigerant. As the refrigerant passes through the outdoor coil, depressurization and evaporation occur, causing it to absorb heat energy. This process can occur even if the outdoor temperature is quite cool. The fluid then passes through pipes (shown as solid lines) to the indoor coil, where the refrigerant undergoes compression and condensation. This causes it to release the heat energy it took in from the outside. The indoor coil becomes considerably warmer than ordinary room temperature. The air is heated up to about 35°C (95°F) as it passes through the indoor coil. This warm air is blown into the ductwork and circulated throughout the house.

Figure 3-6 is a simplified functional diagram of the same heat pump operating to transfer heat energy from the indoor environment to the outdoor environment. This is the "air conditioning mode." Indoor air is drawn or blown by a fan through a coil that contains a refrigerant. As the refrigerant circulates through the indoor coil, depressurization and evaporation occur, so it absorbs heat energy. This chills the air that moves past the coil, and the cooled air is blown into the ductwork, so it circulates throughout the house. The chilling process can also remove some excess humidity from the indoor air, causing the indoor coil to "sweat." The heated fluid passes through pipes (shown as solid lines) to the outdoor coil. In the outdoor coil, the refrigerant undergoes compression and condensation. This causes it to give up the

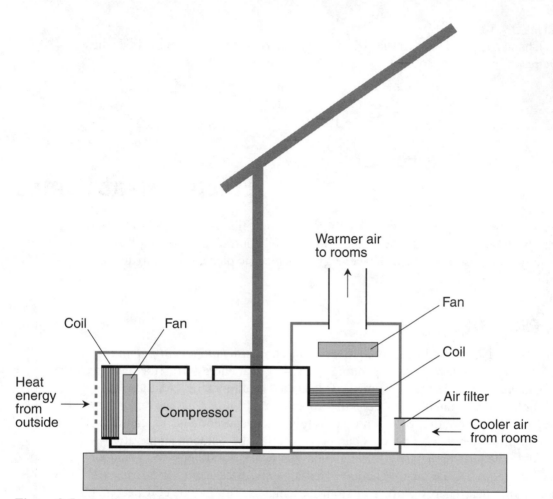

Figure 3-5 An air source heat pump, operating so that it transfers heat energy from the outdoor environment into a building.

heat energy it acquired from indoors. The outdoor fan blows warm air into the environment.

GROUND SOURCE (GEOTHERMAL)

Some heat pumps extract thermal energy from beneath the earth's surface rather than from the outside air, and transfer this energy into a house or building. Figure 3-7 illustrates this principle. Basically, this is a modified air exchange system. The outdoor coil is placed underground, so no fan is necessary. In some systems, the outdoor coil can be placed near the bottom of a deep pond or lake. Heat transfer

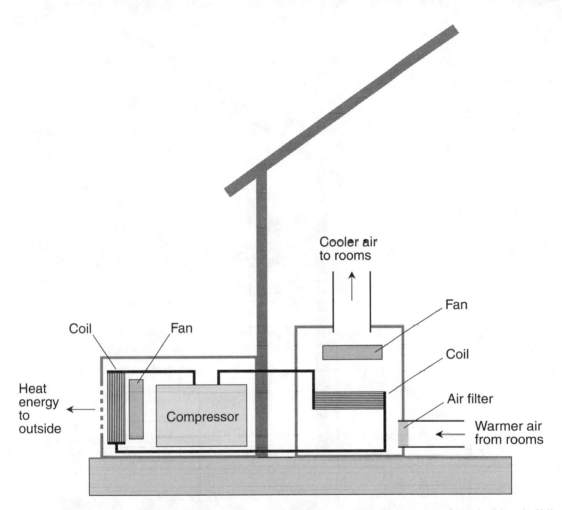

Figure 3-6 An air source heat pump, operating so that it transfers heat energy from inside a building to the outdoor environment.

occurs by conduction from the earth to the coils. This is a *ground source heat pump*, also called a *geothermal heat pump*.

In some locations, the earth is quite warm beneath the surface. Thermopolis and Saratoga, Wyoming, and Hot Springs, South Dakota, are locations in the United States with plenty of available *geothermal heat* despite severe winters. Iceland is another often-cited example. In locations such as these, a ground source heat pump will work at much lower outdoor air temperatures than will an air exchange heat pump. The outdoor coil can even be replaced by a network of pipes, buried deep enough to allow the heating of water that can be directly circulated through the house or building.

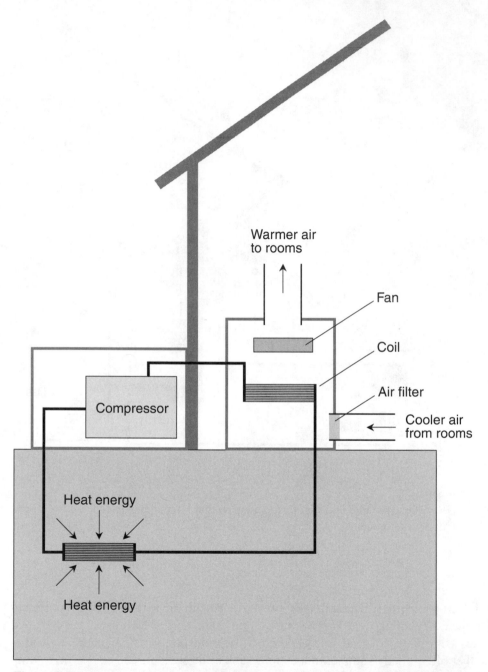

Figure 3-7   A ground source (geothermal) heat pump, operating so that it transfers heat energy from the earth into a building.

Ground source heat pumps can operate to cool the indoor environment during the summer in most locations. However, in some places (such as those mentioned above) the subsurface temperature is high. While this is ideal for home heating with ground source heat pumps, it does not lend itself to interior cooling with that technology.

ADVANTAGES OF HEAT PUMPS

- Heat energy contained in the outdoor environment is a renewable and practically unlimited resource.

- For heat pumps operating in cooling mode, the outdoor environment is a practically infinite heat sink.

- Heat pumps are the most efficient known method of using electricity to heat the indoor environment, provided the outdoor temperature is warmer than approximately 4°C (39°F). They actually deliver more heat energy into the house than they demand from the electric utility.

- Because heat pumps work best when they are called upon to provide a constant indoor temperature, it is not necessary to change the thermostat setting unless you plan to be out of the house for several days.

- The exhaust from a heat pump is either cold or warm air, depending on the mode. There is no CO produced, nor any other noxious gas. (Some pollution is generated, however, at the distant electric power plant if it burns fossil fuels.)

- Ground source heat pumps with sufficiently deep outdoor coil systems can function well even in places where winters are severe. At a depth of several meters beneath the surface, the temperature is constant and is at least 10°C (50°F) in most locations.

LIMITATIONS OF HEAT PUMPS

- Air exchange heat pumps work well if the outdoor temperature is warmer than about 4°C (39°F), but if it gets colder than that, there is not enough thermal energy in the outdoor air to allow efficient operation.

- In older systems that use *chlorofluorocarbon* (CFC) refrigerant compounds, the potential for *ozone depletion* is an issue. A small amount of CFC can destroy large numbers of ozone molecules. Ozone helps to shield the earth's surface from excessive solar UV rays.

- The air that comes from a heat pump is near 35°C (95°F). This is warmer than the typical indoor environment, but it won't heat up a cold house very fast.

- Heat pumps are relatively expensive to install new. This is especially true of the deep ground source type. It may take a long time for a new system to pay for itself.

Quiz

This is an "open book" quiz. You may refer to the text in this chapter. A good score is eight correct. Answers are in the back of the book.

1. To the nearest degree, the lowest possible temperature on the Fahrenheit scale is
   a. −100°F.
   b. −273°F.
   c. −460°F.
   d. There is no limit to how low the Fahrenheit temperature can get.

2. Suppose the difference between the indoor and outdoor temperature is 15°F. That is approximately the same as a difference of
   a. 27°C.
   b. 15°C.
   c. 8.3°C.
   d. 5°C.

3. Which of the following is *not* an advantage of an air exchange heat pump over electric resistance heating?
   a. An air exchange heat pump produces more heat energy per watt hour of electrical energy demanded from the utility, provided the outdoor temperature is warmer than approximately 4°C (39°F).
   b. An air exchange heat pump can work no matter how cold it gets outside, while an electric resistance heater won't work when the outdoor temperature is below freezing.
   c. An electric resistance heater poses little or no risk of electrocution, but an air exchange heat pump is dangerous in this respect.

    d. An air exchange heat pump produces some CO gas, but an electric resistance heater produces none.

4. The number of calories per gram required to warm or cool a substance while it remains in the liquid state is called the

    a. stabilization temperature.

    b. heat of condensation.

    c. heat of vaporization.

    d. None of the above

5. Suppose the difference between the indoor and outdoor temperature is 11 K. That is approximately the same as a difference of

    a. 20°F.

    b. 11°F.

    c. 9°F.

    d. 6°F.

6. If the voltage applied to the heating element in a portable electric space heater is cut in half while the element resistance remains constant, then the heat power the element produces, in British thermal units per hour,

    a. becomes twice as great.

    b. becomes four times as great.

    c. becomes half as great.

    d. becomes one-quarter as great.

7. If the current through the element in an electric space heater doubles while the element resistance remains constant, then the heat power the element produces, in watts,

    a. becomes twice as great.

    b. becomes four times as great.

    c. becomes half as great.

    d. becomes one-quarter as great.

8. When the refrigerant in an air conditioning system condenses, it

    a. releases heat energy to the surrounding environment.

    b. absorbs heat energy from the surrounding environment.

    c. neither releases nor absorbs heat energy to or from the environment.

    d. generates CFC compounds that can damage the ozone layer.

9. Suppose the difference between the indoor and outdoor temperature is 20 K. That is approximately the same as a difference of

a. 36°C.

b. 20°C.

c. 18°C.

d. 11°C.

10. An advantage of a ground source heat pump over an air exchange heat pump is the fact that

a. a ground source heat pump never releases CFC compounds, but an air source device sometimes releases some.

b. a ground source heat pump may occasionally release CFC compounds, but an air source device always releases some.

c. a ground source heat pump can work as an air conditioner (cooling system for the indoor environment), but an air source device cannot.

d. in some locations, a ground source heat pump will work efficiently when it is too cold outside for an air source heat pump to function.

CHAPTER 4

Passive Solar Heating

Solar energy is a practical source of heat for small buildings in locations that receive reasonably abundant sunshine. In the United States, the weather is sunnier in the West than in the East (with the exception of the Pacific Northwest), and the sunlight is more intense in the South than in the North. In South America and Australia, the northern regions get more intense sunlight than the southern regions. Africa straddles the equator and gets good solar radiation all year round.

Sunnyside Glass

Solar energy can be directly harnessed for interior heating by placing large windows on the south side of a building, in conjunction with heat-absorbing and heat-retaining floors, walls, and furniture. Windows can also be placed in steeply pitched roofs, facing generally south. Ideally, the view of the sun from these windows should be unobstructed.

HOW IT WORKS

When the sun shines through windows, visible light and shortwave IR penetrate the glass easily. Dark objects in the room absorb the visible light and shortwave IR energy and re-radiate it in the form of longwave IR. The window glass is less transparent to longwave IR than to visible light and shortwave IR (see Figure 4-1). The longwave rays are reflected from light-colored walls, ceilings, and furniture, "bouncing around" and further heating up dark objects in the room. Longwave IR continues to radiate from dark objects in the room after the sun has gone down or the weather has turned overcast. Unless the objects are massive, this effect does not last long.

We can call the scheme of Figure 4-1 *basic passive solar heating*. It is effective anywhere the sun shines on most days, even if the outside temperature is frigid during part of the year. It works in any room with large south-facing windows. During the night, vertical blinds or heavy curtains can help to minimize the heat loss through the window, which can occur by conduction and convection as a result of contact between the air in the room and the cold window glass.

Basic passive solar heating can work even in climates that most people consider severe in the winter, provided there is enough sunshine. The Bighorn Basin of Wyoming is a good example of such a place. Basic passive solar heating would be less practical in a region such as the coast of Oregon, where, although winter temperatures are not terribly cold, the sky is overcast most of the time.

Most homes can take advantage of sunshine in south-facing windows and, to some extent, windows that face in any direction in the southern half of the compass. Even a small window, with January sun streaming in, can produce considerable

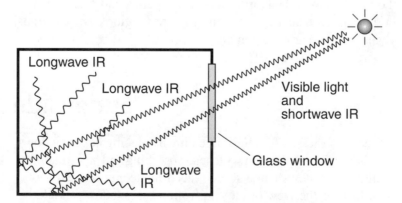

Figure 4-1 When sunlight shines through a window, shortwave IR and visible light penetrate the glass, which is less transparent to the longwave IR radiated from objects in the room. Therefore, heat energy accumulates in the room.

warmth. The houses that work best for basic passive solar heating, in the absence of modification, are those that have a long side facing directly south, with large windows and minimal obstructions. By opening blinds and curtains when the sun shines on the windows, and closing the blinds and curtains when the windows are not exposed to direct sunlight, you can save on your heating bill.

ADVANTAGES OF BASIC PASSIVE SOLAR HEATING

- The expense is minimal. No major structural changes are necessary. A dark couch "throw" or two, some new dark carpeting or dark rugs, and some efficient blinds or curtains are all the hardware that is needed.

- A little personal behavior modification can result in real energy savings, even in the absence of any other changes. Open the blinds or curtains when the sun is shining into exposed windows, and close them the rest of the time.

- Snow on the ground increases the amount of solar energy that enters the windows on sunny winter days, because the snow reflects sunlight. Snow, of course, is most frequent during the heating season.

- Direct sunlight can help to elevate your mood!

LIMITATIONS OF BASIC PASSIVE SOLAR HEATING

- Sunshine on furniture and carpeting can cause color fading over time. Moreover, the effect occurs unevenly, because some portions of a room receive more direct sunlight than others.

- A room may overheat during the day if it receives abundant, direct sunshine for several continuous hours. If your goal is to stay warm at night, you will be tempted to "get ahead of the game," thereby driving the afternoon temperature in some rooms to uncomfortably high levels.

- Some buildings are not "passive solar friendly." These include structures that are shaded by various objects, buildings on north-facing slopes, and buildings in which there are few or no windows facing south. In these cases, passive solar heating may not be a viable option in any form.

- The same windows that allow sunshine in during the day, thereby heating the interior, can lose heat at night unless they are well designed. Thermal insulation such as caulking, along with multipane glass, can help. You must also remember to close the curtains or blinds at night.

PROBLEM 4-1

When I was a child, my mother would close the curtains over the big picture window in the living room during the daylight hours, especially in winter when low-angle sunshine would otherwise fill the room. She feared that the sunlight would fade the furniture, the carpeting, and even the wallpaper. Can this fading be prevented, while still letting all that solar energy come into the house?

SOLUTION 4-1

Fading can be reduced by the use of double-pane or triple-pane glass windows. Most color fading is the result of UV radiation, not IR or visible light, and glass is nearly opaque to UV. However, fading cannot be avoided altogether. Couch "throws" and inexpensive rugs can help, but you will have to find a good balance between esthetics and comfort.

Thermal Mass

Basic passive solar heating can be effective, but if used in the absence of any other heating method, it allows significant fluctuations in temperature between sunny periods and cloudy periods, or between day and night. In order to smooth out these variations, *thermal mass* can be added to, or built into, a structure. This produces *thermal inertia*, so the air in a building tends to remain at a fairly constant temperature despite short-term variations in the solar energy input.

FLOOR AND WALL SLABS

Substances with high material density make the best thermal masses. Concrete is one of the best, and cheapest, materials for this purpose. Brick, adobe, and stone also work well.

Figure 4-2 shows thermal mass in the floor and interior walls of a room that receives direct sunlight through a glass window. In the ideal case, the sun shines directly on the surface of the thermal mass to the greatest extent possible. The surface of the mass should be dark, so it absorbs most of the visible light and shortwave IR energy that strikes it. Bare concrete is unattractive to most people as an interior wall or floor finish, although it can be made to look elegant by special finishing or painting. Interior brick, with concrete behind it, can make a presentable room interior.

When the sun shines into the room (see Figure 4-2A), the thermal mass absorbs energy primarily from visible light and shortwave IR. The thermal mass radiates some longwave IR, which gradually dissipates in the room, warming the walls,

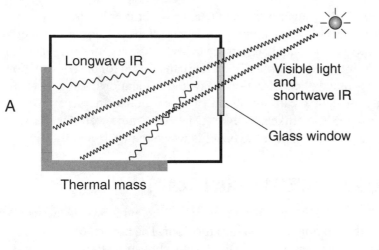

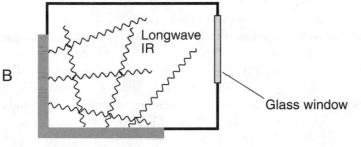

Figure 4-2   Thermal mass in interior walls and floor. When the sun shines (A), the mass absorbs heat energy as visible light and shortwave IR. During cloudy periods or at night (B), the mass re-radiates the heat energy as longwave IR.

floor, ceiling, and furniture, and ultimately the air by conduction. When there is no solar input (see Figure 4-2B), the thermal mass slowly gives up its heat energy in the form of longwave IR. The thermal mass also conducts heat to the air by direct contact.

When a thermal mass is installed in a floor, it must be insulated from the earth below to prevent heat loss to the ground by conduction. Thermal masses should not be used in outside walls if the intent is to heat the building, unless those walls have exposed interior surfaces and substantial insulation is installed between the wall and the exterior surface.

Thermal inertia is proportional to total thermal mass. The best results, if there are long periods with little or no sun, are obtained with large thermal masses. The more thermal mass is installed, the longer it takes to heat up—but it will release heat for a longer time, too. In some passive solar homes, concrete slabs as thick as one meter (1 m) or more are used in floors and subterranean walls to maximize the thermal mass. In the extreme, reinforced and poured concrete can be used for the construction of an entire house! This mode of construction is favored in locations prone to severe hurricanes where flying debris can smash an ordinary house to pieces. As an incidental plus, a house of this sort keeps out external noise quite well. Of course it is expensive, but some people consider the benefits worth the up-front cost.

ROOF WINDOWS AND CEILING OVERLAYS

Passive solar heating can be accomplished by installing windows in a steeply pitched roof, along with a thermal mass on the attic floor. The bottom of the thermal mass forms the ceiling of the living space. Direct sunlight warms the thermal mass (see Figure 4-3A) by irradiating it from the top with visible light and IR. The thermal mass radiates longwave IR downward into the living space by day (see Figure 4-3A) as well as during the night or on cloudy days (see Figure 4-3B).

If basic passive solar heating is used in the living space in addition to thermal mass in the attic floor, the benefits of both methods are combined. An even better scheme involves the use of additional thermal masses in the floor and/or walls of the living space. However, the passive solar heating method shown in Figure 4-3 can function all by itself, even if the living space has few windows. The main challenge is constructing the building so the exterior walls can support the massive attic floor.

PASSIVE SOLAR FORCED-AIR HEATING

Envision a thermal mass in an attic floor, made of concrete blocks laid so their openings are horizontal, and so the openings line up for the entire width of the attic. This creates airways through the thermal mass (see Figure 4-4). Blowers can be used to drive air from the living space, through these airways, and back into the living space.

In a *passive solar forced-air heating* system of this type, the air intake vent, along with the set of blowers, is placed in one of the walls. The air outflow vent is placed in the opposite wall. A second set of blowers can be installed near the outflow vent. A network of ducts ensures that the air passes uniformly through all the airways produced by the openings in the blocks. Heat energy is transferred from the blocks to the air as it flows through. In addition, heat is transferred to the living space by means of longwave IR radiation downward from the thermal mass.

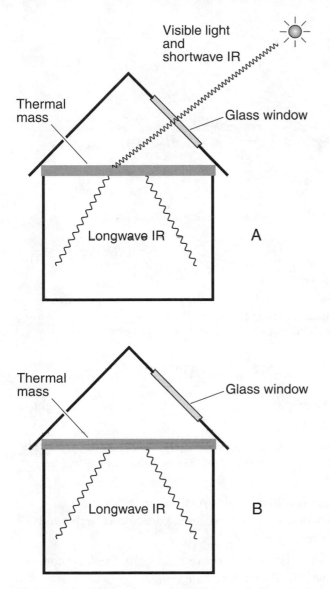

Figure 4-3   Thermal mass in ceilings (attic floors), along with windows on the sunny side of a steeply pitched roof. When the sun shines (A), the mass absorbs energy as visible light and shortwave IR, and also radiates longwave IR into the rooms below. During cloudy periods or at night (B), the mass continues to radiate longwave IR into the living space.

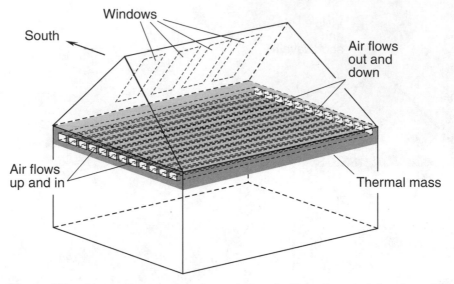

Figure 4-4   Air passageways in a thermal mass facilitate forced-air heating with a passive solar system. In this simplified diagram, the blowers and ductwork are not shown. Outlines represent the perimeters of the interior surfaces.

ADVANTAGES OF THERMAL-MASS HEATING

- Thermal mass reduces fluctuations in temperature between day and night, and between sunny days and cloudy days.

- If there is a prolonged cloudy spell, the heat acquired during sunny periods will last longer in a building with more thermal mass than in a building with less.

- The inclusion of thermal mass in new construction can result in a building more likely to withstand severe weather, particularly high winds accompanied by flying debris.

- A building constructed with substantial concrete, stone, or brick is more fire-resistant than a conventional frame building.

- Thermal mass can provide a measure of acoustic insulation (soundproofing) between the rooms or levels of a house, and between the interior and exterior.

LIMITATIONS OF THERMAL-MASS HEATING

- The cost of constructing a new home with significant thermal mass can be high. Concrete is not cheap, and spot shortages sometimes occur.

- Retrofitting a conventional building with thermal mass can be complicated and expensive. If not done properly, it can also create a danger for the occupants.

- Some people consider thermal mass unattractive, no matter how it is disguised.

- A building with a lot of thermal mass "feels heavy" when you're inside it! Some people like this feeling; others do not.

PROBLEM 4-2
Why doesn't wood work well as a thermal mass? It absorbs and radiates heat, doesn't it? Why should density be important?

SOLUTION 4-2
As density (in kilograms per cubic meter) increases, so does the number of subatomic particles (particularly protons and neutrons) per unit volume, in general. The more "stuff" there is in a certain volume of space, the more heat energy it can retain, and the longer it takes to lose it all. This is why, for example, stones are used in saunas and sweat lodges. Wood is not particularly dense, and therefore does not work very well as a thermal mass.

Solar Water Heating

If you've ever lived near a lake, you know that the temperature of the water remains almost constant from day to day, even if weather conditions vary. This is because water has excellent thermal mass characteristics—superior even to stone, brick, or concrete. Have you noticed that a swimming pool takes a long time to get warm from the sun, and an equally long time to cool down on cloudy days or in cold weather?

HOW IT WORKS

Water can be heated by passing it through pipes embedded in black metal panels under glass, exposed to direct sunlight. These sealed panel-and-glass assemblies are known as *flat-plate collectors* or *flat-panel collectors*. The best place to mount them is flush on a steeply pitched, south-facing roof, well secured to keep wind from breaking them loose. Alternatively, they can be mounted with the flat surface perpendicular to a line running up into the sky toward the noonday, midwinter sun. This can be done in a yard or field, or on a flat or moderately pitched roof. The combination of the *greenhouse effect*, in which longwave IR is trapped beneath the glass, and the fact that the black panels absorb (and convert into heat) all the radiant energy that strikes them can produce panel temperatures higher than the boiling point of water on sunny days, even in the middle of winter.

If the heated water is to be used for washing or bathing, an indirect heating scheme works better than passing the water through flat-plate collectors directly. Special *heat-transfer fluid* allows heat energy to be stored in a hot-water reservoir resembling the tank of a conventional water heater. The heat-transfer fluid is formulated so it won't freeze in the panels during frigid overcast or nighttime conditions. The freezing point can be sufficiently lowered by adding ethanol (ethyl alcohol) to ordinary water, or by using antifreeze similar to the mixtures used in cars and trucks. The storage tank is large (at least several hundred liters or 100 gallons), giving its contents substantial thermal mass. The tank is well insulated, so it does not give up its heat energy until hot water is called for. Figure 4-5 is a simplified diagram of a water heating system that uses flat-plate collectors. Theoretically this is an active system considered as a whole, because of the presence of the pump for the heat-transfer fluid, which operates on electricity.

If the energy derived from flat-plate collectors is to be used for heating the home interior, no water or storage tank is necessary. The heat-transfer fluid can be passed directly through the collectors and pumped through pipes embedded in concrete thermal masses in the floors or walls of the building. In this case, the flat-plate collectors take the place of a gas-, propane-, or oil-fired boiler, and the heat-transfer fluid takes the place of the water. The circulation pump is shut off during the night and on cloudy days, but the thermal mass continues to radiate heat into the living space.

ADVANTAGES OF SOLAR WATER HEATING

- The system requires minimal external power. Only the pump requires electricity, and it does not consume much power.

- It is a closed system, neither leaking nor requiring the introduction of outside substances.

- It generates no pollutants, except indirectly as a result of electricity used by the pump (if the power plant burns fossil fuels).

- This type of system makes almost no noise. The pump motor is small and quiet, and can be located in the basement or other out-of-the-way place.

- There are not many moving parts to break down.

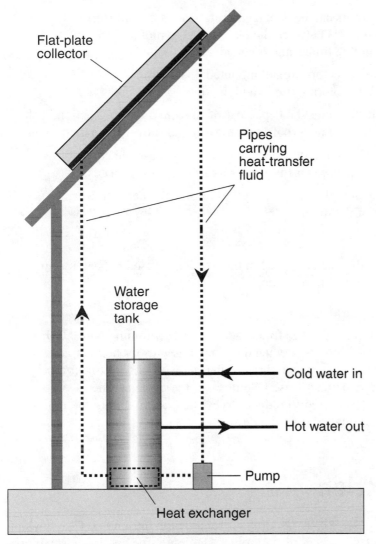

Figure 4-5 Simplified diagram of a water heating system that uses flat-plate collectors and heat-transfer fluid.

LIMITATIONS OF SOLAR WATER HEATING

• This type of system will not work for prolonged cloudy periods if the weather is extremely cold.

- Flat-plate collectors can be damaged or destroyed by hail storms, ice buildup (if water gets between the plates and the roof surface and then freezes), falling tree limbs, and other adverse events.

- If the flat-plate collectors are not mounted flush with a roof or other surface, they can be torn loose in a high wind.

- Flat-plate collectors must be kept clear of snow in order to function. If the collectors are located on a rooftop, removing the snow can be inconvenient.

- Systems for heating a living space require many flat-plate collectors, driving up the expense. This is especially true in cold climates.

- Flat-plate collectors do not work well in locations where the weather is overcast most of the time.

PROBLEM 4-3

Can a system such as the one described in this section be used to heat a swimming pool? It seems like a logical thing to do.

SOLUTION 4-3

Flat-plate collectors have been used for decades to heat swimming pools. If the pool is located outdoors and is not used when the temperature drops below freezing, the pool water can be directly circulated through flat-plate collectors. In extremely hot locations such as the American Desert Southwest, the cycle can be reversed. The flat-plate collectors can be used to radiate excess heat away during the night, and the system can be shut down during the day.

Sidehill Construction

An ideally designed passive solar building is "square with the compass" so the long sides of the house face directly north and south. The south side has the most window area, and the north side has the least. (In the southern hemisphere this is reversed; the north side has the most window area, and the south side has the least.) The north side is well insulated, so it loses minimal heat energy. All of these criteria taken together have given rise to the concept of *sidehill construction*, in which the structure is built into the earth on the south face of a hill. This tactic was used by pioneers in the American Great Plains when they built "soddies" partway into the sunny sides of whatever small hills they could find.

HOW IT WORKS

Figure 4-6 is a simplified drawing of a passive solar home that uses sidehill design. Windows in the south side of the pitched, asymmetrical roof produce heat by greenhouse effect during the daylight hours. At night, automatically controlled blinds close the roof windows, minimizing heat loss. Windows on the south side of the living space admit sunshine during the day, and curtains or manual blinds cover these windows at night. Thermal masses in the attic floor (the ceiling of the main living space), as well as in the north wall and the floor, facilitate heat retention so the home remains warm at night and on cloudy days. Foam or fiberglass insulation is provided behind and beneath the wall and floor thermal masses, minimizing conductive heat loss into the earth.

In Figure 4-6, the east and west walls of the house are not shown because the view is looking straight east through the house. However, it is apparent from this drawing that both the east and the west walls are about half below the surface and

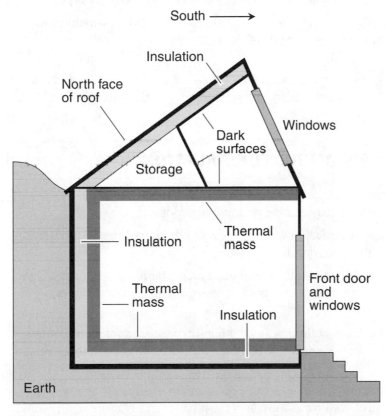

Figure 4-6   An example of sidehill building design with thermal masses and passive solar windows. This is a cross-sectional view looking directly east.

half above the surface. These walls can also have thermal mass along with exterior insulation, similar to the design of the north wall. Windows can be included to admit some of the morning and afternoon sunshine into the living space. These windows, like all of the windows in this house, should have curtains or blinds that are closed at night.

ADVANTAGES OF SIDEHILL CONSTRUCTION

- The heat retention is excellent if the insulation is good. (If the insulation is inadequate, however, this type of home will lose excessive heat to the earth.)

- This type of home is protected from cold north winds, and to some extent from the winds that occur during violent thunderstorms in the temperate latitudes.

- This type of building lets in less external noise than most other types.

- There is less exterior surface area, which means less exterior maintenance work, compared with other types of homes.

- A home of this sort stays cooler in summer than a conventional home. The windows can be opened at night. During daylight hours, the windows can be closed and the blinds drawn on the side(s) of the house exposed to sunshine.

LIMITATIONS OF SIDEHILL CONSTRUCTION

- Earth movement can be a problem in some locations. Over time, an unstable hill will destroy a home built into the slope.

- Proper drainage is critical in order to prevent flooding of the interior during spring snowmelt or heavy rain storms.

- Walled-off rooms in the north side of this type of building will receive little or no natural light unless translucent windows are placed in the interior walls.

- The roof of a sidehill building can be a liability issue if someone ventures onto the roof from the north side and then falls off the east, south, or west side.

- The accumulation of *radon gas* can be a problem in any part of a building that is below ground. Proper measures must be taken to ensure that radon levels do not become unhealthful in the living space.

PROBLEM 4-4
The view toward the north is fabulous from my proposed home site. Even though the slope is uphill looking north, there are trees in the yard, and the snow-capped peaks in the distance can be seen, especially in winter when the leaves are off the trees. I want to see these things from inside the house, but still take advantage of sidehill design. How can I do this?

SOLUTION 4-4
Small windows can be installed in the north side of a house with sidehill design. The earth does not have to come all the way up to the level of the roof. The windows must be near the ceiling, but that's all right because the view is looking upslope. If tinted plastic is added to the surfaces of these windows by a professional installer, multiple-pane glass is used, and efficient blinds or curtains are employed to minimize radiation loss at night, you can still enjoy that northerly view during daylight hours.

Quiz

This is an "open book" quiz. You may refer to the text in this chapter. A good score is eight correct. Answers are in the back of the book.

1. In a basic passive solar heating system, which of the following can increase the amount of solar radiation entering a room, if all other factors remain constant?

   a. Leaves on deciduous trees

   b. Snow on the roof

   c. Snow on the ground

   d. The installation of thermal masses

2. Heat loss by radiation from a house at night can be minimized by

   a. a building design with a steeply pitched roof.

   b. orienting the house so the long sides face north and south.

   c. automated window blinds that close at night and open during the day.

   d. the use of dark furniture and carpeting.

3. Fill in the blank in the following sentence to make it true: "When there is no solar radiation, a thermal mass slowly gives up its heat energy as _____, keeping the living space warm."

   a. ultraviolet radiation and convection in the walls

   b. infrared radiation and conduction to the air

   c. forced air through insulation in the walls

   d. conduction to the outside earth

4. Which of the following types of solar radiation have the most difficulty penetrating a pane of glass?

   a. Visible light

   b. Shortwave IR

   c. Longwave IR

   d. All three of the above types of solar radiation penetrate glass equally well.

5. A good position in which to mount a flat-plate collector is

   a. on the south side of a steeply pitched roof.

   b. on a flat roof, facing straight up.

   c. in a vertical position, facing due north.

   d. in a vertical position, facing due south.

6. A sidehill construction plan should be ruled out if a home is to be built

   a. where earth movement is known to occur.

   b. where the wind blows hard and often.

   c. where the sun shines almost every day in the summer.

   d. where the winters are exceptionally cold.

7. Which of the following contributes to heating a room in a passive solar arrangement?

   a. Visible light

   b. Shortwave IR

   c. Longwave IR

   d. All of the above

8. Basic passive solar heating without significant thermal mass

   a. cannot, in a practical sense, provide any net heating effect.

   b. may tempt the user to overheat sunny rooms during the day.

   c. requires sidehill construction in order to work properly.

   d. requires a steeply pitched roof in order to work properly.

9. Which of the following substances is the least effective at retaining heat energy?

   a. Wood

   b. Water

   c. Stone

   d. Concrete

10. Which of the following is a practical advantage of adding ethanol to the heat-transfer water in a solar water heating system?

   a. Ethanol prevents bacteria, viruses, and other pathogens from growing in the heat-transfer water.

   b. Ethanol keeps the heat-transfer water from freezing if the outdoor temperature becomes extremely low.

   c. Ethanol prevents oxidation in the heat-transfer pipes, prolonging their life.

   d. Forget it! Ethanol should never be added to heat-transfer fluid because the resulting mixture is flammable.

CHAPTER 5

Exotic Indoor Climate-Control Methods

This chapter outlines a few of the less common ways of heating and cooling a building. The schemes described here are not intended as primary modes of indoor environment control, but to offload some of the burden from conventional systems.

Direct Wind-Powered Climate Control

A *wind turbine* can be connected to an electric generator, which in turn can be connected to electric heating elements. With proper voltage regulation, this can provide supplemental heat for a modest-sized home whenever the wind blows at a reasonable speed.

HOW IT WORKS

Figure 5-1 is a block diagram of a wind turbine, equipped with an electric generator, connected into a zone-type electric baseboard heating system. The voltage is

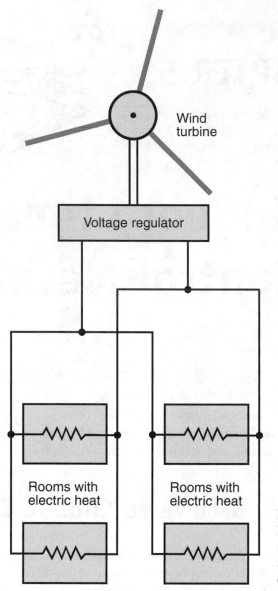

Figure 5-1 A supplemental interior heating system that uses a wind turbine and voltage regulator connected into a conventional electric home heating circuit.

regulated to remain at or below 117 V AC, so the heating elements are not damaged by excessive current.

A typical wind turbine designed for residential use can produce 12 kW of power on a day with moderate wind. That is the equivalent of eight electric space heaters, each rated at 1500 W. From Chapter 1, recall that 1 kW = 3410 Btu/h. That means the wind turbine system can provide approximately $3410 \times 12 = 40,920$ Btu/h. (Let's round this off to 40,000 Btu/h). A gas furnace for a small or medium-sized home produces about 80,000 Btu/h when running "full blast." Thus, on a bitter-cold winter day with some wind, the system shown in Figure 5-1 can theoretically supply half of the energy demand to keep a small or medium-sized house warm.

A system like that shown in Figure 5-1 is dependent on wind for its operation. The large amounts of energy required for home heating cannot be stored in batteries at a cost affordable to most homeowners. In a location where the wind does not blow often, this scheme will not prove cost effective, no matter how long the time frame. However, in some places, winters are cold and windy. Places like that are good candidates for a system such as the one diagrammed in Figure 5-1.

On a hot, windy day, a wind-powered system can be used to operate an air conditioner or evaporative cooler. Searing dry winds, ironically, tend to occur in the same places as deadly cold blizzards. But if the weather is hot and calm, some other source of energy will be necessary to cool a home.

Wind power is discussed in greater detail later in this book. For now, it is worth noting that the wind can be an excellent source of supplemental or intermittent electrical power, but wind does not lend itself well to use as a continuous, high-volume energy source. When properly exploited, wind energy can significantly reduce dependence on fossil fuels. For the idealist, that is good news. But it would be a bad idea to disconnect or uninstall an existing gas, propane, or oil furnace in favor of a system like that shown in Figure 5-1.

ADVANTAGES OF DIRECT WIND-POWERED CLIMATE CONTROL

- It can significantly reduce (but not eliminate) reliance on conventional methods of home heating.

- It will function in some situations when heat is in demand but the normal home heating system won't work because of a loss of electric utility power.

- It is a nonpolluting system. It generates no greenhouse gases, particulates, carbon monoxide, ground contamination, or waste products.

- The electricity produced by the wind turbine can be used for other intermittent purposes, such as the charging of automotive batteries and laptop computers. When heat is not needed, the electricity can be used to provide interior lighting and to operate other devices that are not sensitive to sudden voltage drops.

- The wind turbine can, if desired, later be used for a stand-alone or interactive wind energy system. Such systems are described later in this book.

LIMITATIONS OF DIRECT WIND-POWERED CLIMATE CONTROL

- The wind is an intermittent source of energy. When the wind is not blowing at least a few miles per hour, a direct wind-powered system will not operate. This type of system is, therefore, not practical in locations where the average wind speed is low.
- Wind turbines will not work properly if the wind is too strong.
- A small wind turbine can be wrecked by a powerful thunderstorm, hurricane, or ice storm.
- A system such as the one described here will take a long time to pay for itself, and in fact may never. Wind turbines are expensive. The motivation must go beyond simple bottom-line economics.
- Some people may not like the appearance of a wind turbine. In addition, it will make some noise when it is running. Residential wind turbines are seldom seen in cities or suburbs for these reasons.

PROBLEM 5-1
Will a conventional, old-fashioned windmill operate as a wind turbine? When I drive around in the country, I occasionally see these. Some of them appear to be in good physical condition.

SOLUTION 5-1
A few people have adapted traditional windmills, used on farms in years gone by, to generate electricity. Their original function was to pump water from a well. However, in any application where intermittent voltage reductions can be tolerated, and where a continuous source of power is not needed, a windmill can be connected to an electric generator and voltage regulator, and the result will be useful energy for supplemental heating. But old-fashioned windmills are not as efficient as modern wind turbines.

Direct Hydroelectric Climate Control

A *water turbine* can be connected to an electric generator, which in turn can be connected to electric heating and cooling systems. With sufficient water flow and proper voltage regulation, this can provide most or all of the heating and cooling that a typical home needs.

HOW IT WORKS

Figure 5-2 is a block diagram of a small water-driven energy system adapted for electric baseboard home heating. This is essentially the same as the direct wind-

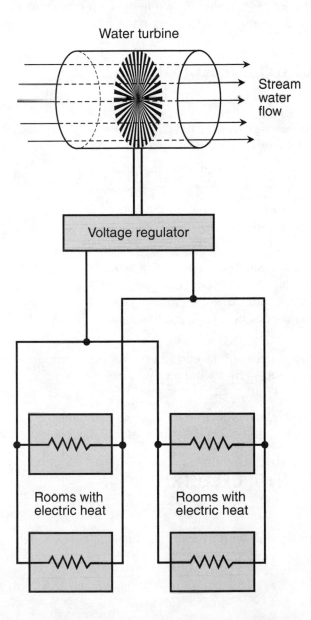

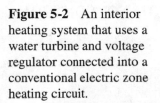

Figure 5-2   An interior heating system that uses a water turbine and voltage regulator connected into a conventional electric zone heating circuit.

powered system shown in Figure 5-1, except that the wind turbine is replaced by a water turbine. As with the wind system, the voltage is regulated to remain at or below 117 V AC. The system can also be used to power electric air conditioners.

A good water turbine, installed in a fast-moving stream or small river with sufficient vertical drop, can produce 20 kW of power on a reliable basis. Again, recall that 1 kW = 3410 Btu/h. Therefore, a substantial water turbine system can provide approximately $3410 \times 20 = 68,200$ Btu/h. (Let's round this off to 70,000 Btu/h). That can just about keep a small or medium-sized home comfortable in all types of weather, as long as the stream or river doesn't dry up or freeze solid.

ADVANTAGES OF DIRECT HYDROELECTRIC CLIMATE CONTROL

- Direct hydroelectric power can significantly reduce, and possibly eliminate, reliance on conventional methods of home heating and cooling.

- Water flow, unlike wind, is continuous, as long as the stream or river is large and fast enough.

- A direct hydroelectric system will function when the normal indoor environmental control system won't work because of a loss of electric utility power.

- This technology is virtually nonpolluting. It generates no greenhouse gases, particulates, carbon monoxide, ground contamination, or waste products. A small amount of heat is imparted to the stream water as a result of friction with the water turbine components, but this is rarely significant in a small-scale system.

- The electricity produced by a water turbine can be used for other purposes, in a manner similar to that provided by a wind turbine—but on a more reliable basis.

- A water turbine can, if desired, be modified to serve in a stand-alone or interactive hydroelectric system of the sort described later in this book.

LIMITATIONS OF DIRECT HYDROELECTRIC CLIMATE CONTROL

- Only a lucky few live on properties with rivers or streams that have enough flow to provide significant hydroelectric power.

- A small stream may periodically dry up in a prolonged drought, or freeze solid (from surface to bottom) in an extreme cold snap. Then, obviously, a water turbine won't work.

- A water turbine requires considerable water mass, along with a significant vertical drop, in order to provide enough power to heat a home. This may necessitate the installation of a small dam, which could give rise to environmental and regulatory issues.

- A system such as the one described here will take a long time to pay for itself, and in fact may never. A home hydroelectric system is comparable in cost to a wind-powered system that provides the same amount of usable output.

PROBLEM 5-2

There is a stream running through my property. It is about 10 m (33 ft) wide and 2 m (6 ft) deep in the middle. It flows well except in the winter, when it freezes on the surface, although the water keeps flowing under the ice. The land is nearly flat. The vertical drop, from the point where the stream enters the property to the point where it exits, is only 0.5 m (approximately 18 in). Will this stream provide enough hydroelectric power to heat my home?

SOLUTION 5-2

You can get an engineer to evaluate the situation, but it sounds like your stream won't provide enough power to heat a home. Even so, a water turbine in this stream could produce a few watts of electricity on a reliable basis to operate small appliances.

Direct Photovoltaic Climate Control

Arrays of *photovoltaic* (PV) *cells* are normally used in conjunction with storage batteries, or else as a supplement to the electric utility. But such arrays, called *solar panels*, can be used in a stand-alone system without batteries if the user is willing to accept the fact that when sunlight is insufficient, the system will not operate.

HOW IT WORKS

Figure 5-3 is a simplified block diagram of a direct PV system for indoor environment modification. In bright sunshine, a single *silicon PV cell* produces approximately

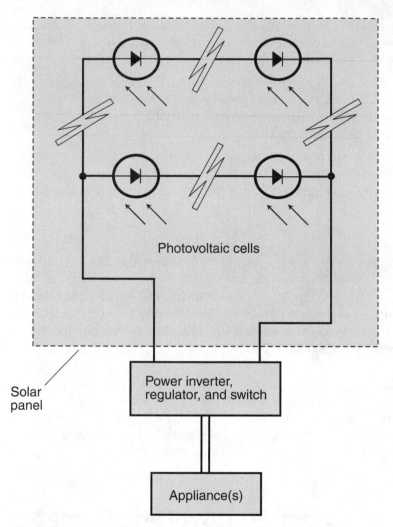

Photovoltaic cells

Solar
panel

Power inverter,
regulator, and switch

Appliance(s)

Figure 5-3 A supplemental indoor environment control system that uses a solar panel, power inverter, voltage regulator, and switch connected to a small appliance such as a fan or humidifier.

0.5 direct-current volts (V DC). They are connected in an enormous *series-parallel array* that provides 12 V DC or 24 V DC output at fairly high current in direct sunlight. An array of this type is known as a *solar panel*.

The output of the solar panel goes to a circuit called a *power inverter*. This changes the DC output of the solar panel into 117 V AC that can be used by ordinary home appliances. There may be a *voltage regulator* to ensure that the voltage remains fairly constant under conditions of varying solar intensity. There is also an *automatic shutdown switch* that disconnects the solar panel and turns off the system if the illumination becomes too dim to properly operate appliances connected to it.

Electric space heaters and air conditioners draw too much current to be powered by a solar panel of reasonable size. (Theoretically, such appliances can be powered this way, but the benefit would rarely, if ever, justify the cost.) Electric floor fans, ceiling fans, humidifiers, and evaporative coolers do not draw much current, and they can be used intermittently with a direct PV source of power. In summer in a place such as southern Arizona, for example, the hottest part of the day is usually attended by bright sunshine. In that sort of location, the system shown in Figure 5-3 could operate a set of ceiling fans in a home or business. In a cold but sunny desert region such as northern Nevada or central Wyoming, the system could operate a humidifier to mitigate the extreme dry indoor conditions that prevail in winter.

ADVANTAGES OF DIRECT PV CLIMATE CONTROL

- A direct PV system has no batteries and no complex tie-in to the electric utility. This minimizes the cost.

- Because the system is simple, there isn't much that can go wrong with it, provided the installation is done properly, and as long as reasonable care is taken to protect the solar panels from damage.

- There is practically no maintenance involved with this type of system. Once it is up and running, it can be pretty much left alone.

- A direct PV system can later be upgraded to serve as a stand-alone PV power system (using storage batteries) or an interactive PV power system (tied into the electric utility).

LIMITATIONS OF DIRECT PV CLIMATE CONTROL

- A direct PV system will only operate when there is enough illumination. This means that the sun must be shining, or it must be a fairly bright cloudy day.

- In the event of a hot day without enough illumination to operate the system, or during a hot night, the system won't work.

- A hail storm or high wind can ruin a set of solar panels in a hurry.

- If the solar panels get covered with snow, the snow must be removed manually in order for the system to function.

- Care must be taken to ensure that the current demanded from the system is well below the maximum deliverable current. If the delivered current momentarily drops (as will occur on a partly cloudy day when a cloud passes in front of the sun), the switch will shut the system down for that time if it is operating near peak capacity.

PROBLEM 5-3

I have plenty of real estate, and can place a solar panel array of practically unlimited size on it. Money is no object here! Can I run an air conditioner or electric space heater using the direct PV scheme shown in Figure 5-3 if I install a huge PV array on the property?

SOLUTION 5-3

Yes. But if your funds are unlimited, you might consider a hybrid system that makes use of wind power as well as solar power. Small-scale stand-alone and interactive photovoltaic and wind-powered systems are discussed later in this book.

Thermal-Mass Cooling

The thermal inertia that enables dense solids such as concrete, brick, and stone to keep a home warm can be also be exploited to keep an interior living space cool. This is done in regions where summer heat presents more of a problem than winter cold.

HOW IT WORKS

Figure 5-4 is a simplified cross-sectional view of a ground-floor room in a home that uses thermal mass for cooling. The mass is under the floor and on the exterior wall. There is no insulation between the floor mass and the earth beneath, so heat energy can dissipate into the ground by conduction. The exterior walls are painted white, or covered with white siding, to minimize the heat they absorb from direct sunlight.

If the room shown in Figure 5-4 faces in a direction where the sun regularly shines, the window can be closed during the daytime (A) and opened at night (B). This keeps the hot outdoor air from entering during daylight hours, but allows warm indoor air to escape at night. A set of curtains or blinds can be drawn whenever the sun would otherwise shine into the room. When the window is not exposed to sunlight, the curtains or blinds can be opened. If the room happens to face north (or south in the Southern Hemisphere), then the curtains or blinds play a less important role, although they may have to be drawn in the early morning and late afternoon hours in summer when the sun is in the poleward half of the sky.

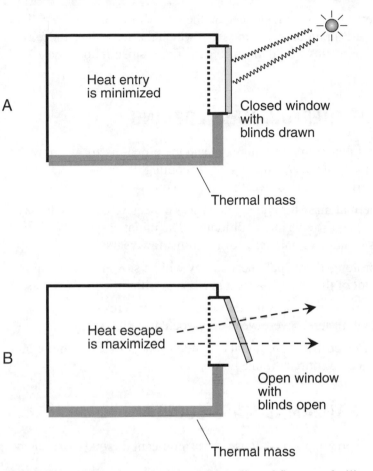

Figure 5-4 Thermal masses in exterior walls and floor can facilitate cooling. In daylight (A), the mass absorbs heat energy from the air in the room. At night (B), heat energy from the room escapes through the window.

The thermal mass on an exterior wall is slow to heat up during the day, and is also slow to cool down at night. Therefore, it stays near the average air temperature during the course of the day. In the desert, where the difference between afternoon and predawn temperatures can exceed 25°C (45°F), this average is considerably below the peak afternoon temperature. A person who is used to "meat-freezer-mode" air conditioning might find this a little uncomfortable in the height of the summer season, but it is an improvement over the conditions that would exist in a conventional frame house without air conditioning. (Arguably, it is also healthier than overdriven air conditioning that produces an extreme contrast between outdoor and indoor temperatures.)

The ideal geometry for an arrangement of the sort shown in Figure 5-4 is a sidehill home that opens up in the direction from which the sun shines the *least*. In the Northern Hemisphere, this would mean constructing a sidehill home on a north-facing slope.

ADVANTAGES OF THERMAL-MASS COOLING

- The proper installation of thermal mass, and the proper use of windows and curtains or blinds, can keep a home from becoming dangerously hot, no matter how torrid the outdoor weather may become.

- The use of thermal mass for cooling is far less expensive on a day-to-day basis than the use of active air conditioning. The initial cost outlay for the thermal mass installation may pay for itself in a few years.

- If there is a prolonged hot spell, heat energy will be slower to build up in a house with a lot of thermal mass than in a house with little thermal mass.

- The inclusion of thermal mass in new construction can result in a house more likely to withstand severe weather, particularly high winds.

- A house constructed with substantial concrete, stone, or brick is more fire-resistant than a conventional frame house.

LIMITATIONS OF THERMAL-MASS COOLING

- The cost of constructing a new house with significant thermal mass can be high.

- Retrofitting a conventional building with thermal mass can be complicated and expensive, and if not done right, potentially dangerous.

- Some people consider thermal mass unattractive, no matter how it is disguised. (However, others enjoy it.)

- In a home built into the poleward-facing slope of a hill, and with few or no windows that admit direct sunlight, the relative lack of natural light can be a problem for some people.

PROBLEM 5-4

If the exterior of a home is painted white, or covered with white siding, won't that impede the radiation of heat away from the building at night, thus reducing the effectiveness of the thermal mass in the exterior wall? Is there any solution to this?

SOLUTION 5-4

A white exterior surface does impede the radiation of heat at night. But this is more than offset by the fact that it also slows down the absorption of heat when the sun shines on it. In the hottest part of the year at temperate latitudes, daylight takes up a much greater part of the day than darkness. Perhaps someone will invent a paint that is normally dark or black, but that turns white when exposed to sunlight. If applied directly to the exposed, outside surface of a thermal mass that forms an exterior wall, this would enhance the radiation of heat away from the building at night, while impeding the absorption of heat during the day!

Evaporative Cooling

When water changes from a liquid to a gaseous state, it absorbs heat energy, as you have learned. It isn't necessary for water to actively boil in order for this to occur. The effect takes place at any temperature where water is normally a liquid, and this range extends well above and below the normal human comfort zone. By accelerating the process of water evaporation, a body of liquid water can be made to remove heat from the air. This is the principle behind *evaporative cooling*.

HOW IT WORKS

When hot, dry air passes through a water-soaked foam or fiber *filter* several millimeters thick, some of the water is forced to evaporate. As a result, the air emerging from the filter is several degrees cooler than the air going in. The water for the filter is replenished from reservoirs that receive a constant supply of fresh (ideally distilled) liquid water.

Figure 5-5 is a simplified functional diagram of an evaporative cooler designed to be placed on a flat, level rooftop. The assembly is cube-shaped, measuring approximately 1 m (3 ft) high, wide, and tall. Water from the reservoirs is constantly absorbed by the filters, which are specifically designed for maximum water uptake by *capillary action*. The water is held at a constant level in the reservoirs by a floating-ball device similar to the ones used in toilet tanks. A large *squirrel-cage fan* pulls hot outside air in through the filters. The hot air causes some of the water in the filters to evaporate, so the air emerging from the filters is cooler than the outside air. The cooled, somewhat more humid air is blown into the living space. The outlet is a single vent in the center of the ceiling of the room chosen for cooling. This sets up convective currents in which cool air descends in the middle of the room and moves radially outward near the floor. Air leaves the room by vents near the floor.

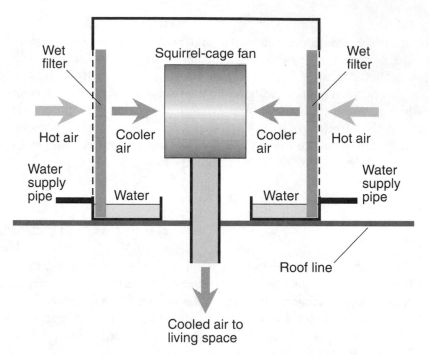

Figure 5-5 Simplified functional diagram of a roof-mounted evaporative cooler. This type of system functions best in a hot desert environment.

The result is a constant supply of fresh, cool air for the living space (assuming the outdoor air is not polluted).

The system shown in Figure 5-5 works best in a single-level building, and ideally it is placed in the room most often used by the largest number of people. The output of an evaporative cooler can be blown into a network of ducts and vented to multiple rooms. But that diminishes its cooling effectiveness, because the ducts can impart heat back into the air.

An evaporative cooling system works best when the air is particularly hot and dry. It will work to some extent even if the outdoor temperature is only a little above the comfort level. The poorest performance occurs when the outdoor air is humid. Evaporation is limited when the humidity is high, as compared to when it is low. Evaporative coolers are most popular in hot desert regions such as southern California, Arizona, and New Mexico. These systems are also commonly used in the Australian Outback.

ADVANTAGES OF EVAPORATIVE COOLING

- An evaporative cooling system uses about 20 percent of the electrical energy consumed by a conventional air conditioner to provide the same amount of cooling. The only significant current is drawn by the blower. Over time, a lot of money can be saved as a result of the lower electric bills.

- Because evaporative systems consume minimal electricity, they cause less greenhouse gas production or other pollution from the utility power plants.

- Evaporative cooling systems are simpler than refrigeration-type units. This means there is less to go wrong! If the unit does malfunction, repair costs are usually low.

- Evaporative cooling systems do not use ozone-depleting compounds, as some older refrigeration-type systems do.

- Outside air is constantly supplied to the living space. This eliminates "stuffiness" that can sometimes occur with recirculation in conventional air conditioning systems.

- The filters help to get rid of dust and particulate pollution in the outdoor air.

- An evaporative system will not produce the "meat freezer" effect, in which people entering a building from the outside "hit a wall of cold air."

- Evaporative systems provide humidification in desert locations. This can mitigate physical complaints associated with extreme low humidity. A conventional system, in contrast, drives the moisture content of the air down further.

LIMITATIONS OF EVAPORATIVE COOLING

- An evaporative cooling system won't work well if the humidity is high, because the outdoor air entering the system is nearly saturated and cannot accept much moisture from the filters.

- In a large building with many rooms, a single evaporative cooling unit can serve only one room or zone. Multiple units, one for each room or zone, along with well-insulated ducts, can provide some cooling in such situations, but this drives up the cost and increases the system complexity.

- An evaporative system may not provide enough cooling for comfort if the outdoor temperature becomes exceptionally high, as can occur in locations such as the Mojave Desert in southern California during the summer.

- If an evaporative cooling system leaks water, damage can occur to a flat roof and the interior ceiling below if the roof does not have proper drainage.

- The filters must be periodically cleaned or replaced. If the environment contains a lot of dust or other particulate matter, this maintenance job will have to be performed often.

PROBLEM 5-5

My house has a steeply pitched roof. How can I install an evaporative cooler up there?

SOLUTION 5-5

A support structure can be built, with a level platform on which the cooling unit can rest. This platform can be placed at the peak of the roof. A straight duct can be installed below the blower to let the air pass into the living space through the attic. The platform structure must be sturdy so it can support the weight of the unit and the water it contains, and so a high wind won't rip the whole thing loose. The duct must be insulated so the cooled air does not get hot again as it passes through the attic.

Subterranean Living

The ultimate insulation from heat and cold is achieved by placing the entire living structure underground. This can be done so that the ceiling of the highest level is located at the surface. Then the roof doubles as a yard or driveway. *Subterranean living* can also be carried on in structures far beneath the surface, such as in defunct mines or missile silos!

HOW IT WORKS

The temperature beneath the earth's surface remains remarkably constant, once a certain depth (a few meters) is achieved. This is true even in locales where the difference between the highest and lowest temperatures during a typical year can exceed 70°C (126°F). In most temperate regions, the subterranean temperature at a depth of a few meters is around 10°C (50°F). At greater depths, one encounters progressively warmer constant temperatures.

Subterranean homes must be constructed of concrete and steel. This allows them to withstand the tremendous pressures that exist at even modest depths. Of course, there can be no direct windows to the outside, although "periscope windows" can be installed. A complex ventilation system is required, and some means must be put in place to ensure a constant flow of air. The structure must incorporate shock-absorbing insulation that prevents vibrations in the earth, such as are caused by motorized traffic or other activity at the surface, from being transmitted to the living space.

As with any type of housing, subterranean living works better in some places than in others. The best locations are where the risk of earthquakes is minimal, and there are few or no underground caverns, streams, or veins of magma (molten rock that produces volcanoes). A subterranean dwelling or structure must be located entirely above the water table.

ADVANTAGES OF SUBTERRANEAN LIVING

- Subterranean living can eliminate heating and cooling costs almost entirely.

- An underground house is storm-proof. Even a tornado will not do any damage to the main structure, although surface-mounted ventilation duct openings, solar panels, wind turbines, or other peripheral apparatus will be destroyed.

- Subterranean living is practical in congested areas, even in the downtowns of large cities. This has been realized, for example, in a portion of Tokyo, Japan, in the form of a project called *Geotropolis*.

- With proper acoustical design, an underground home in a city can be totally soundproofed against noise pollution from the world above. There will be no problems with sirens, barking dogs, loud parties, and other sources of "audio aggravation."

- Living underground may provide psychological comfort for people afraid of global catastrophes such as all-out nuclear war, however unlikely such events may be.

LIMITATIONS OF SUBTERRANEAN LIVING

- Underground construction is not a good idea in an earthquake zone, unless massive structural reinforcements and shock absorbers are put in place.

- Underground buildings do not last long where there is slow, but unstoppable, earth movement. This is common in more places than most people suspect. It is essential to check the history of a location, and to scrutinize existing structures (especially older ones) for signs of settling.

- Subterranean living can cause claustrophobia in susceptible individuals. Some people can't tolerate being underground, even for a few minutes.

- You won't have any direct panoramic views of the outdoors from an underground house. You'll have to settle for large, well-illuminated paintings, murals, or photographs.

- Radon gas can be a problem, unless exceptional ventilation and proper construction techniques are used.

- Ventilation is important in any underground dwelling. Otherwise the oxygen content of the air will drop as inhabitants breathe and rebreathe the air.

- An underground dwelling is impractical where the water table is high. The entire structure must be above the highest level reached by the water table during prolonged wet periods.

PROBLEM 5-6

How can natural light be obtained in abundance in a subterranean home? Wouldn't a massive opening at the surface, along with a huge array of mirrors, be impractical, as well as degrading the thermal insulation? I am building an underground house, but am concerned about the possible negative health effects of daylight deprivation.

SOLUTION 5-6

Electric lamps have been developed that have an output spectrum (intensity as a function of wavelength) similar to that of direct sunlight. These light sources can be placed at desired locations throughout the home. And of course, one can always spend a lot of time outdoors, doing all of the things surface dwellers do, such as walking, hiking, cycling, and commuting.

Quiz

This is an "open book" quiz. You may refer to the text in this chapter. A good score is eight correct. Answers are in the back of the book.

1. Which of the following statements concerning a direct hydroelectric home heating system is false?

   a. It will likely take a long time for the initial cost of a direct hydroelectric home heating system to be recovered.

   b. Relatively few people can take advantage of this technology because they don't have access to the necessary streams or rivers.

   c. If the water turbine is properly designed, this type of system can function even if the stream or river freezes solid from the surface to the bottom.

   d. A system of this type produces no greenhouse gases or carbon monoxide.

2. Which of the following summer cooling technology combinations would be the cheapest to operate in a hot desert?

   a. A conventional air conditioner operated by the electric utility

   b. A sidehill home with a southern exposure and thermal mass in the interior walls

   c. An evaporative cooler powered by a solar panel and a power inverter

   d. An air source heat pump operated by the electric utility

3. The voltage regulator in a direct wind-powered heating system

   a. shuts it down when the wind is not blowing.

   b. allows it to work even when the wind is not blowing.

   c. prevents excessive current from flowing through the heaters.

   d. keeps the turbine from spinning too fast.

4. The widespread adoption of evaporative cooling systems as replacements for conventional air conditioners would be expected to reduce greenhouse gas pollution because

   a. evaporative systems do not use refrigerant compounds, which can contribute to the destruction of the earth's ozone layer.

   b. evaporative systems demand less energy from power plants, many of which burn fossil fuels and thereby release CO_2.

   c. evaporative cooling systems release heat energy into the atmosphere, which tends to increase the rate at which greenhouse gases are absorbed or changed into other, harmless, compounds.

   d. Forget it! Greenhouse gas pollution would increase if evaporative cooling systems were adopted on a large scale to replace conventional air conditioners.

5. In direct sunlight, a single silicon PV cell produces approximately

   a. 0.5 V DC

   b. 1.5 V DC

   c. 12 V DC

   d. 24 V DC

6. The lack of a backup storage battery or utility tie-in with a direct wind-powered home heating system

    a. means that it will not work when the wind is not blowing.

    b. allows it to serve the electric needs of a typical household all the time.

    c. means that it cannot save the user any money on electric bills.

    d. will cause the utility company to charge you extra when you are using it.

7. If you live in an underground house and a tornado passes directly over your property, you should check immediately after the storm to see if

    a. damage has occurred to the thermal mass in the walls, floor, or ceiling.

    b. damage has occurred to the external ventilation intakes and outlets.

    c. earth movement has taken place.

    d. Forget it! You need not have the slightest concern about tornadoes if you live underground.

8. A solar panel of reasonable size, along with a power inverter, cannot operate a resistance-type electric home heating system because

    a. a power inverter cannot provide sufficient voltage.

    b. a solar panel produces unregulated voltage.

    c. a solar panel delivers too much voltage.

    d. a solar panel of reasonable size cannot deliver enough current.

9. Fill in the blank to make the following sentence correct: "A sidehill home with external thermal mass, built for optimum summer cooling in a desert in the Southern Hemisphere, would ideally be built on a _____ slope."

    a. north-facing

    b. south-facing

    c. east-facing

    d. west-facing

10. A device that changes the DC output of a solar panel to standard AC utility electricity is known as a

    a. backup generator.

    b. power inverter.

    c. series-parallel array.

    d. voltage regulator.

CHAPTER 6

Conventional Propulsion

Conventional propulsion systems include machines powered by gasoline, diesel fuel, jet aircraft fuel, and rocket fuel. All of these fuels work by means of *combustion*—controlled explosions, actually—in which reactive substances yield *kinetic energy* along with various chemical compounds. Gasoline- and diesel-powered vehicles are sometimes called *fossil-fuel vehicles* (FFVs).

Gasoline Motor Vehicles

Automobiles, trucks, trains, boats, propeller-driven aircraft, and various other machines "burn" *gasoline* in *spark-ignition engines*, which are specialized *internal-combustion engines*. In the United Kingdom, gasoline is known as *petrol* (pronounced "PET-rel").

WHAT IS GASOLINE?

Gasoline is an aromatic, flammable liquid that consists mainly of compounds called *hydrocarbons*. We've seen some examples of such compounds: methane, heating oil, and propane. Gasoline is a chemical relative of these. In a theoretically perfect engine, one liter of gasoline yields approximately 3.48×10^7 J of kinetic energy. If a U.S. gallon (1 gal) of gasoline is burned outright, it yields approximately 1.25×10^5 Btu of heat.

Gasoline is an extremely *volatile* substance, which means that it readily changes state from liquid to vapor at room temperature. Gasoline vaporizes at an increasing rate as the temperature increases. For this reason, different grades of gasoline are produced for different geographical regions, and for different seasons of the year. In general, more volatile gasoline is used in winter and in cold locations, and less volatile gasoline is used in summer and in warm locations.

Gasoline is produced in oil refineries, along with other fossil fuel products. Constituents of gasoline include compounds known as *alkanes*, *alkenes*, and *cycloalkanes* in proportions that depend on the nature of the crude oil available and the grade of gasoline produced. The highest grades of gasoline contain the highest amounts of *octane*, a hydrocarbon that helps prevent *engine knock*. This phenomenon can occur when gasoline ignites too soon in an internal-combustion engine.

Some grades of gasoline contain approximately 10 percent added *ethanol*, also called *ethyl alcohol*, which is derived from corn or other grain. This gasoline is sometimes called *gasohol*. It is not the same thing as *E85*, which is 85 percent ethanol and only 15 percent gasoline.

Besides hydrocarbon atoms, gasoline contains *additives* intended to improve the efficiency with which it burns. In the middle of the 20th century, *tetraethyl lead* was a common additive. When it became known that lead is toxic to humans and animals and that it tends to accumulate in the environment, efforts were begun to discontinue the use of tetraethyl lead. Nowadays, "leaded fuel" is uncommon. Other additives vary from country to country, and also as new technologies and government regulations evolve.

HOW A GASOLINE ENGINE WORKS

A gasoline engine takes advantage of the flammable nature of gasoline to generate mechanical energy. The basic concept is shown, greatly simplified, in the functional diagram of Figure 6-1. The gasoline is mixed with air by a *carburetor* to form a fine mist of gasoline droplets. This mixture is injected into an enclosed *cylinder* containing a movable *piston*. An *electric arc*, produced by a *spark plug*, ignites the gasoline/air mixture, causing a small explosion that makes the vapors in the cylinder violently expand, driving the piston downward. This spark and explosion together

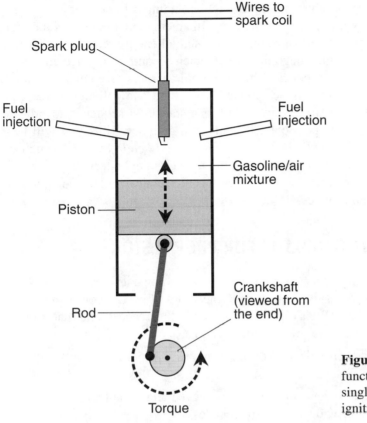

Figure 6-1 Simplified functional diagram of a single cylinder in a spark-ignition gasoline engine.

are called *ignition*. The piston is connected to a rotatable *crankshaft* by a rod and bearings. The rotation of the crankshaft carries the rod around because the whole assembly has a certain amount of *angular momentum*. Thus, the piston slides back upward in the cylinder after it has reached its lowest position. Each single upward or downward sweep of the piston is called a *stroke*.

When the piston nears the top of its up-and-down motion cycle, the gasoline/air mixture is injected into the cylinder again, and another spark is ignited at just the right moment. In a *two-stroke engine*, ignition takes place every time the piston reaches the top of its cycle, so there are two strokes (one down, one up) of the piston for every ignition event. In a *four-stroke engine*, ignition takes place at the top of every other cycle, so there are four strokes (one down, one up, another down, and another up) for every ignition event.

If the gasoline/air *injection* and the spark ignition occur repeatedly and rhythmically, and if everything is timed exactly right, the crankshaft rotates with enough *torque* to turn a *gear drive* or *belt drive* connected to a wheel, propeller, electric generator shaft, or other rotating assembly. Timing is important in order for the engine to operate with optimum efficiency. In an engine, *efficiency* is the ratio of the actual kinetic energy output to the total potential energy required to operate the engine. Improper timing degrades the efficiency, and can place increased physical strain on engine components. In the extreme, it can cause engine failure.

Most internal-combustion engines contain more than one cylinder of the type shown in Figure 6-1. The strokes are timed in a staggered sequence to provide smoother operation than is possible with only one cylinder. Thus you will hear about *two-cylinder*, *four-cylinder*, *six-cylinder*, and *eight-cylinder* engines. Sometimes you'll encounter engines with more than eight cylinders. In general, as the number of cylinders increases, so does the mechanical energy an internal-combustion engine can produce.

ADVANTAGES OF GASOLINE FOR PROPULSION

- Gasoline has a high *energy-to-mass ratio*, also known as *energy density*. In other words, for a given quantity of fuel, a lot of useful work can be obtained. Few other fuels can rival gasoline in this respect, and that is one of the reasons why it is still in widespread use.

- Gasoline engines provide a lot of power in proportion to their bulk and weight.

- A well-manufactured gasoline engine is rugged and reliable. If it is well cared for, it can provide service for decades.

- A well-maintained and tuned gasoline engine can operate over a wide range of temperatures, humidity levels, and barometric pressure levels— encompassing almost all conditions encountered on this planet.

LIMITATIONS OF GASOLINE FOR PROPULSION

- Gasoline combustion, even when complete, produces carbon dioxide (CO_2), a known greenhouse gas.

- When gasoline does not burn completely (as is the case in any real-world engine), some carbon monoxide gas (CO) is produced. This gas is deadly if it leaks into a vehicle, or if a gasoline engine is run in an enclosed area.

- Gasoline presents a fire hazard if it is not properly stored, or if it leaks.

- Sulfur compounds in some forms of gasoline can contribute to acid rain.

- Gasoline engines, when used in machines such as "leaf blowers" or "weed whackers," can cause annoying *noise pollution*.

- Some of the compounds in gasoline are known to increase the risk of cancer in humans and animals directly exposed to them over a period of time.
- Gasoline is refined from crude oil, the price of which can vary suddenly and dramatically.
- Gasoline is derived from crude oil, a nonrenewable resource.

PROBLEM 6-1

Haven't emission-control devices and regulations practically eliminated the pollution problems caused by gasoline engines?

SOLUTION 6-1

Emission reductions afforded by technology and legislation have been largely offset by the increased number of gasoline engines in use throughout the world, especially in countries with rapidly developing economies. Emission-control technology isn't perfect; it does not get rid of all the pollutants produced by gasoline engines. The problem with CO_2 has, in fact, been getting worse in recent years, and the prognosis is not good. This gas is not toxic, but it has been implicated as a possible contributor to global warming.

Petroleum Diesel Motor Vehicles

Conventional *diesel fuel*, also called *petroleum diesel* or *petrodiesel*, is used in applications similar to those in which gasoline is used. Diesel engines are employed more often than gasoline engines to propel large trucks, locomotives, agricultural implements, and other heavy vehicles.

WHAT IS DIESEL FUEL?

Diesel fuel is derived from petroleum in the oil refining process. It is a hydrocarbon mixture. In these respects, it resembles gasoline. But diesel fuel is more energy-dense than gasoline. In a theoretically perfect engine, a liter of petroleum diesel can provide approximately 4.09×10^7 J of kinetic energy. If 1 gal of petroleum diesel is set on fire, it yields approximately 1.47×10^5 Btu of heat.

Because of the greater energy density of diesel fuel compared with gasoline, a diesel engine can travel farther on a given volume of fuel than can a gasoline-powered engine of the same size pulling the same mass over the same terrain. Diesel fuel is also more dense than gasoline in the physical sense. For a given volume, diesel fuel masses about 18 percent more than gasoline. Some engineers consider diesel fuel to bear a closer resemblance to heating oil than to gasoline.

Some diesel fuels contain vegetable oils and/or liquefied animal fats along with petroleum-derived hydrocarbons. The proportion of *biological oil* to petroleum fuel can vary over a wide range. Biologically derived diesel fuel is known as *biodiesel*. The addition of biodiesel to petroleum diesel reduces the emission of sulfur compounds when the fuel is consumed; the biological product contains less sulfur than the petroleum product. This may be significant if the use of biodiesel becomes widespread, because sulfur emission has been implicated as a cause of acid precipitation that is known to disrupt ecosystems.

HOW A DIESEL ENGINE WORKS

A diesel engine is a compression-ignition system that generates mechanical energy by means of combustion, just as is the case with a gasoline engine. The major difference is that a diesel engine does not have spark plugs or the electrical system that spark plugs require.

Figure 6-2 is a simplified functional diagram of a cylinder in a diesel engine. The process is nearly identical to that which occurs in a gasoline-engine cylinder. As the piston moves upward, the heat produced by compression ignites the fuel/air mixture. When the piston nears the top of its up-and-down motion cycle, the fuel/air mixture is injected into the cylinder, so that combustion can take place.

As with a gasoline engine, timing is important in order for a diesel engine to operate with optimum efficiency and minimum pollutant emissions. Diesel engines contain multiple cylinders, as do gasoline engines. As the number of cylinders increases, so does the mechanical energy the engine can produce.

ADVANTAGES OF PETROLEUM DIESEL FOR PROPULSION

- Diesel fuel is more energy-dense than gasoline.
- In some locations, diesel fuel is cheaper per liter or per gallon than gasoline (as of this writing).
- In heavy vehicles, diesel fuel is preferred because it can deliver more power (actual mechanical energy per unit time).
- A diesel engine is more efficient than a gasoline engine of the same size.
- The combustion of diesel fuel produces less CO gas than the combustion of an equal amount of gasoline.

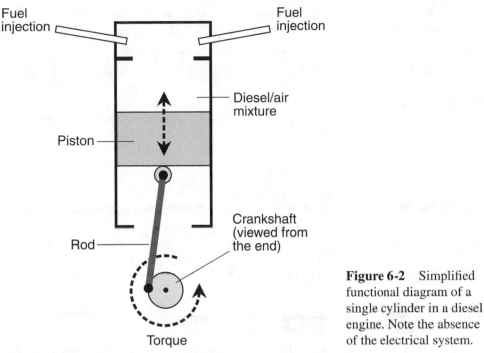

Figure 6-2   Simplified functional diagram of a single cylinder in a diesel engine. Note the absence of the electrical system.

- Diesel engines are more reliable than gasoline engines because diesel engines do not require electrical systems. They're simpler, so fewer things can go wrong!

LIMITATIONS OF PETROLEUM DIESEL FOR PROPULSION

- Diesel contains more sulfur than gasoline, producing higher emissions of sulfur dioxide and other sulfur pollutants.

- A diesel engine is heavier than a gasoline engine capable of the same maximum power.

- The exhaust from diesel fuel has an odor that some people find intolerable.

- When a diesel engine is not optimally tuned, black soot can be emitted in the exhaust, especially when the engine is under full load. This soot contains unburned carbon. This can cause health problems for people in cities where large numbers of diesel engines are in operation.

- A diesel engine is harder to start and keep running than a gasoline engine, particularly in cold weather when diesel fuel thickens into a gel-like substance. It may even crystallize. When that happens, the fuel injector cannot effectively deliver fuel into the cylinders.

- Some of the compounds in petrodiesel are known to increase the risk of cancer in humans and animals directly exposed to them over a period of time.

PROBLEM 6-2
How can the cold-weather problems with diesel fuel be overcome?

SOLUTION 6-2
In some vehicles, the fuel lines, fuel filter, engine block, and cylinders are electrically heated. The source of electricity can be an alternator while the vehicle is running, and external utility power when the vehicle is sitting idle in cold weather.

Conventional Jet Propulsion

Propeller-driven aircraft, such as private and small commuter planes, burn high-octane gasoline. Commercial and military aircraft are mostly jet-propelled, and they require a different sort of fuel. Some aircraft use turbine-driven propeller engines known as *turboprops*. Jet and turboprop engines can burn a mixture of high-octane gasoline and jet fuel.

WHAT IS JET FUEL?

Most jet fuel is based on a compound known as *kerosene*, which is obtained in the oil refining process along with gasoline, petroleum diesel, and other products. Kerosene can also be obtained from coal. This was done as early as the mid-1800s to obtain fuel for *gas lamps* that provided indoor and outdoor illumination before *electric lamps* existed.

Kerosene has various nonaircraft uses. In Japan, it is used for home heating. Some portable stoves, often used by campers and mountain climbers, also employ this fuel. Kerosene produces a spectacular flame when burned in open air, and has been used in the entertainment industry for this reason. It works well as a general solvent, and some people claim that it will remove head lice (although it inflames the skin)! General-purpose kerosene is less refined than the compound that forms the basis of jet fuel. It has a characteristic odor, similar to that of diesel fuel, that can give some people nausea or headaches.

When kerosene is refined for use as aircraft fuel, it is processed to reduce the sulfur content, and also to minimize its natural corrosive properties. The most common kerosene-derived jet fuel used in the United States is called *JET A*. It freezes at −40°C (−40°F) and has an *autoignition temperature* of approximately 425°C (800°F). Some other jet fuels, notably *JET B*, freeze at lower temperatures, but they are more volatile, and are used only in high-altitude flight or polar flight where extremely cold air temperatures exist.

Kerosene-derived jet fuels contain additives such as *antioxidants* (to keep the fuel from becoming too viscous), *electrostatic dissipation* substances (to prevent "static electricity" from causing sparks and consequent fires or explosions), chemicals to reduce the corrosive nature of pure kerosene, *icing inhibitors* to prevent fuel-line freezing, and tetraethyl lead, the compound that was once common as an anti-knock agent in automotive gasoline.

HOW A JET ENGINE WORKS

Figure 6-3 is a functional illustration of a basic jet engine. This is not intended as a literal rendition. It is greatly simplified, and the proportions are exaggerated in some cases and minimized in others to show the interaction among the basic components.

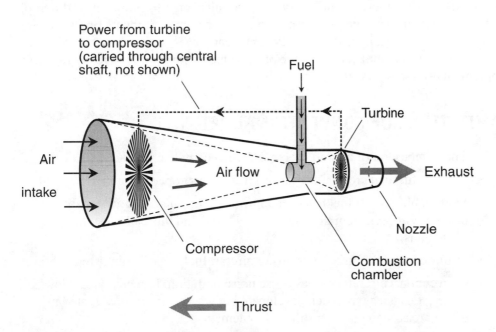

Figure 6-3   Simplified functional diagram of a conventional jet aircraft engine.

Air enters through a large opening and is driven into the engine by a *compressor*, also called a *fan*. The compressed air passes down the length of the engine into the *combustion chamber*, where fuel is injected and the air-fuel combination is burned. The drastic increase in the temperature produces extreme pressure in the chamber, and hot gases emerge from the rear of the combustion chamber at high speed. A turbine in this gas stream provides power for the compressor once the engine is running. (Initially, the compressor must be started by means of an external power source. In the earliest jet engines, power for the compressor was derived from an external piston engine similar to those used in propeller aircraft.) The exhaust leaves the rear of the engine through the *nozzle*. If the speed of the exhaust is higher than the forward airspeed of the whole assembly, *thrust* occurs.

Jet engines work best at high airspeeds (more than approximately 640 km/h or 400 mi/h) because they accelerate small amounts of air by a large factor. For airspeeds less than that, propeller-driven aircraft often perform better because they accelerate large amounts of air by a small factor. Turboprops also work well for aircraft at slower speeds. Turboprops are replacing propeller engines for small and medium-sized commuter aircraft.

ADVANTAGES OF CONVENTIONAL JET FUEL

As of this writing, there is one overwhelming advantage to propeller, turboprop, and jet engines in aviation: no alternatives can compete with them! Engineers are looking into fuel cells, hydrogen, and even nuclear power as alternative methods for providing aircraft propulsion, but none of these technologies is near ready for mass implementation today.

LIMITATIONS OF CONVENTIONAL JET FUEL

- The combustion of aircraft fuel, even when complete, produces CO_2.
- Aircraft fuel can create a fire hazard if it is not properly stored.
- Flammable aircraft fuel has been known to gain entry to aircraft cabins or otherwise leak from the tanks or fuel lines, and this can cause fires and explosions.
- Sulfur compounds in some forms of aircraft fuel can contribute to acid rain.
- Conventional aircraft engines cause noise pollution. In recent years, low-noise jet engines have been developed, but where airports are located in urban areas, the noise can still be problematic.

- Jet engines can produce *contrails* (vapor trails) that may affect the natural dynamics of the upper atmosphere.

- Some of the compounds in aircraft fuel are known to increase the risk of cancer in humans and animals directly exposed to them over a period of time.

- Aircraft fuel is refined from crude oil, and suffers from all the market-related problems that go along with that energy source.

- The world's supply of available crude, and thus of conventional aircraft fuel, will eventually run out. The only question is when.

PROBLEM 6-3

Suppose that hydrogen becomes widely available someday, replacing natural gas for home heating, as well as gasoline and diesel for cars, trucks, boats, and trains. Won't it work well in aircraft, too?

SOLUTION 6-3

Yes, but only if a cost-effective and safe way is found to obtain, transport, and store usable hydrogen fuel in the necessary amounts.

Conventional Rocket Propulsion

Today's rocket fuel falls into two categories: *liquid rocket fuel* and *solid rocket fuel*. A liquid-fueled rocket engine works like a gigantic jet engine, except the oxidant supply is carried along with the vehicle rather than supplied by external air. A solid-fueled rocket engine is significantly different from a jet engine in its basic design. In a sense, it resembles a huge, controlled-combustion firecracker containing an oxidant mixed in with the powder.

WHAT IS LIQUID ROCKET FUEL?

Liquid rocket fuel consists of a *propellant* and *oxidizer*. The propellant in a rocket engine is the counterpart of the fuel in a jet engine, and the oxidizer in a rocket engine is the counterpart of the air that lets the fuel in a jet engine burn. Common rocket-engine propellants are kerosene, hydrogen (liquefied for storage in onboard tanks), and a nitrogen/hydrogen compound called *hydrazine* (N_2H_4). In the case of kerosene and hydrogen propellants, oxygen (liquefied for storage in onboard tanks) forms the oxidizer. This liquefied oxygen is sometimes symbolized by the acronym

LOX. When the propellant is hydrazine, the oxidant is a nitrogen/oxygen compound called *nitrogen tetroxide* (N_2O_4).

The cleanest-burning liquid rocket fuel is hydrogen, which, when combined with oxygen, yields only energy and water vapor. When kerosene is refined for use as a rocket fuel, there are few impurities, but some CO and CO_2 gases are an inevitable byproduct of the combustion of this fuel because of the carbon atoms in its molecules. Hydrazine and nitrogen tetroxide yield considerable quantities of nitrogen when they react. This gas is not toxic, and in fact makes up nearly three-quarters of the composition of the earth's atmosphere at the surface.

WHAT IS SOLID ROCKET FUEL?

The earliest solid rocket fuels were similar to gunpowder, and were used in fireworks and weaponry. Today, this type of fuel is used in model rockets. A typical model-rocket engine is a small cylinder of packed gunpowder-like material about the size of an adult's index finger. It is ignited by a hot wire and burns for a second or two. The thrust provided by a little engine like this can propel a small rocket (about 0.5 m or 18 in tall) to an altitude of several hundred meters (perhaps 2000 ft) if the vehicle is allowed to coast after the fuel is expended.

Basic solid fuel contains a propellant, an oxidizer, and a *catalyst* that facilitates steady, reliable combustion following ignition. These fuel constituents are all originally in powdered form. They are mixed and packed uniformly to ensure an even, sustained, timed burn. A typical military solid-fueled rocket engine contains charcoal (carbon) as the fuel, potassium nitrate as the oxidizer, and sulfur as the catalyst. This combination is called *black powder*. Alternative materials that can be used to make solid rocket fuel include sodium chlorate, potassium chlorate, powdered magnesium, or powdered aluminum. Combinations of these substances are called *white powder*.

GETTING INTO SPACE

Neither black nor white powder can provide enough energy to accelerate a vehicle to *orbital velocity*. That requires a speed of about 29,000 km/h (18,000 mi/h). Advanced solid fuels can produce sufficient thrust in the *booster* (first stage) of a multistage rocket to allow subsequent liquid-fueled stages to put a payload into orbit, or even achieve *escape velocity* for interplanetary travel. The minimum speed required to escape the gravitational influence of the earth is approximately 40,000 km/h (25,000 mi/h).

HOW A LIQUID-FUELED ROCKET ENGINE WORKS

Figure 6-4 illustrates the principle behind a liquid-fueled rocket engine. The propellant and the oxidizer are allowed to flow into the combustion chamber, where an initial ignition event starts the combustion process. As long as the propellant and the oxidizer both keep flowing into the chamber, combustion continues. The depressurization, along with the heating, converts the liquid propellant and oxidizer to a hot gaseous mixture.

Forward thrust results from the *action/reaction* principle: the hot gases, escaping from the nozzle, produce a powerful backward jet. This causes the rocket to accelerate forward. While the rocket is in the atmosphere, forward thrust requires that the escaping gases attain a speed at least equal to the forward speed of the vehicle. However, once the vehicle is in outer space, this is not necessary. Any escaping gas, regardless of its speed, will then impart a force on the vehicle that causes it to accelerate in a forward direction.

HOW A SOLID-FUELED ROCKET ENGINE WORKS

Figure 6-5 illustrates a solid-fueled rocket engine of the sort used in model rockets. (In larger vehicles, the principle is the same as that shown here, although the engine geometry may be somewhat different.) Once ignition occurs, the fuel burns at a

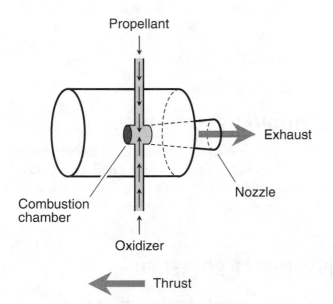

Figure 6-4   Simplified functional diagram of a liquid-fueled rocket engine.

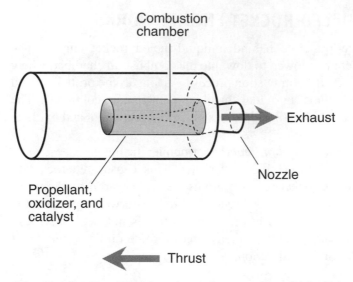

Figure 6-5 Simplified functional diagram of a solid-fueled rocket engine.

controlled rate, producing thrust as the hot-gas combustion byproducts are forced out of the rear opening or nozzle.

Once a solid-fueled engine has been ignited, combustion occurs until all the fuel is exhausted. There is no way to throttle such an engine down or stop the "burn" until it runs its course. This can be a disadvantage of solid-fueled engines as compared with liquid-fueled engines. However, because solid fuel is generally used only for the initial stages of space-bound vehicles, and not for the final stage, this is not a significant problem in practice.

ADVANTAGES OF CONVENTIONAL ROCKET FUEL

Conventional rocket fuels—liquid or solid—provide the only methods for spacecraft propulsion at the time of this writing. Alternative technologies such as *ion engines*, controlled *nuclear fusion* reactions, and *solar sails* are in the conceptual stages, but these vehicle types are not likely to be deployed until near the end of the 21st century.

LIMITATIONS OF CONVENTIONAL ROCKET FUEL

* With conventional rockets, it is impossible to attain speeds much greater than the escape velocity from earth, that is, 40,000 km/h or 25,000 mi/h.

The amount of fuel required to produce high speeds would be so large as to render the vehicle too massive for acceleration to those speeds.

• Conventional rockets are not an efficient method of propulsion in an overall sense. Putting even a small payload into earth orbit requires an enormous amount of fuel—many times the mass of the payload itself.

• Conventional rockets are not practical for interstellar travel because of their limited maximum speed. The nearest star system is about 40 trillion kilometers (4.0×10^{13} km), or 25 trillion miles (2.5×10^{13} mi), away from our solar system. Even at 1,000,000 km/h (620,000 mi/h), a journey to that star system would take more than 4500 years!

PROBLEM 6-4

Can a conventional rocket use a combination of liquid and solid materials to obtain its propulsion?

SOLUTION 6-4

This has been done in the form of *hybrid fuel* with a solid propellant and a fluid or gaseous oxidizer. The trouble is that the solid propellant, without the oxidizer mixed in uniformly, may burn unevenly or incompletely. Nevertheless, *SpaceShipOne*, the first private manned rocket, was a hybrid. The propellant was a solidified substance called *hydroxy-terminated polybutadiene* (HTPB), and the oxidizer was *nitrous oxide*.

Quiz

This is an "open book" quiz. You may refer to the text in this chapter. A good score is eight correct. Answers are in the back of the book.

1. When kerosene is used as the propellant in liquid rocket fuel,

    a. some carbon compounds are inevitably contained in the exhaust.

    b. nitrogen tetroxide must be used as the oxidizer.

    c. it must be mixed with gasoline to prevent excessively rapid combustion.

    d. All of the above

2. Tetraethyl lead is no longer used as an anti-knock additive in gasoline because

    a. it is expensive and it never worked well anyhow.

    b. the lead is toxic and it accumulates in the environment.

    c. it is excessively corrosive and can damage fuel lines and engines.

    d. it is excessively volatile and works only at low temperatures.

3. Which of the following statements about conventional jet aircraft fuel and engines is false?

    a. Some of the compounds in conventional jet fuel can contribute to cancer in humans.

    b. The availability and price of jet fuel is dependent on the supply of crude oil.

    c. Nuclear jet fuel is cheaper and less polluting than conventional jet fuel.

    d. Conventional jet aircraft produce contrails in the upper atmosphere.

4. Petroleum diesel fuel

    a. is more energy-dense than gasoline.

    b. contains more vegetable oil than gasoline.

    c. produces more CO gas than gasoline when burned.

    d. produces odor-free exhaust when burned.

5. Hydrazine is used as

    a. a rocket engine propellant.

    b. a jet engine catalyst.

    c. a rocket fuel oxidizer additive to prevent knocking.

    d. a diesel additive to prevent ozone depletion.

6. Which of the following terms most precisely defines a substance that changes state easily from liquid to vapor at room temperature?

    a. Reactive

    b. Catalytic

    c. Volatile

    d. Flammable

7. Which of the following statements about propeller-type aircraft engines is true?

   a. Propeller engines accelerate less air than jet engines.

   b. Propeller engines accelerate air to a greater extent than jet engines.

   c. Propeller engines work better at high speed than jet engines.

   d. Propeller engines burn a different sort of fuel than jet engines.

8. In an internal-combustion engine, a stroke is

   a. a complete rotation of the shaft.

   b. a single up or down piston movement.

   c. premature ignition of the fuel.

   d. a failure of ignition in a cylinder.

9. The fuel/air mixture in a conventional gasoline engine is ignited by

   a. compression during the upward piston stroke.

   b. decompression during the downward piston stroke.

   c. an electrical spark in the cylinder.

   d. a small pilot light in the cylinder.

10. An important purpose of the turbine in a jet engine is to

   a. provide additional thrust.

   b. filter out noxious gases from the exhaust.

   c. draw the air into the intake opening.

   d. provide power to the compressor.

CHAPTER 7

Propulsion with Methane, Propane, and Biofuels

Various combustible liquids and gases have been tested as substitutes for gasoline and petroleum diesel. A few have been put to practical use. In this chapter, you will learn about some workable alternative fuels for surface propulsion.

Methane for Propulsion

Methane is known mainly as a fuel for interior heating systems, but it can also be used as an alternative to gasoline. The first methane-powered cars were developed before gasoline was widely available. Like the early steam-powered cars, methane vehicles were superseded by powerful, affordable, and mass-produced fossil-fuel vehicles early in the 20th century. Methane for use in car, truck, and boat engines is generally obtained from the earth by the same means as is done for home heating.

But other methods have been developed, most notably the fermentation or composting of plant and animal waste. Methane produced in this manner is sometimes called *biogas*. For propulsion, methane works in essentially the same way as propane, which is discussed in the next section.

ADVANTAGES OF METHANE FOR PROPULSION

- Methane-powered engines are more efficient, in general, than gasoline engines in converting the potential energy into usable mechanical energy.
- The widespread production and use of methane may mitigate problems, such as variable prices and supply interruptions, that occur with gasoline and petroleum diesel.
- Increased production of biogas for use in vehicles would, as a side benefit, increase the methane supply available for heating homes and businesses and for running electric power plants.
- In many countries, including the United States, distribution pipelines for commercially produced methane already exist. It would not be difficult or expensive to set up a vast network of refueling stations for methane-powered vehicles.
- Biogas can be produced in small-scale composting plants. The supply does not have to come exclusively from centralized sources. This could enhance the security of the civilized world by distributing energy resources and assets, making them less vulnerable to natural or human-caused disasters.

LIMITATIONS OF METHANE FOR PROPULSION

- In its combustible form, and at ordinary room temperature and pressure, methane is a gas. This presents handling and transportation problems, as with any industrial gas.
- While methane-powered engines are efficient, they do not offer the *high performance* obtained in the best gasoline-powered engines. If you enjoy auto racing (as a participant), methane may not be the fuel of choice for you.
- Methane was not widely available at vehicle refueling stations in most countries at the time of this writing. This included the United States.
- The production of biogas by composting can produce objectionable odors. There is also some concern that the process, if not done responsibly, could lead to the breeding and spread of disease-causing microorganisms.

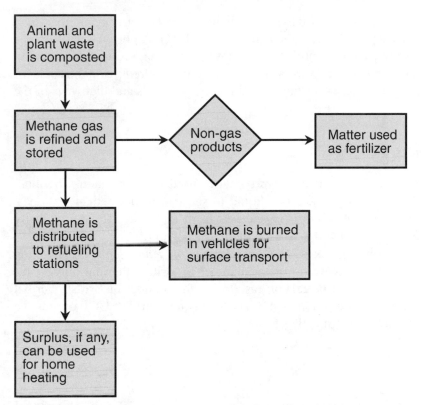

Figure 7-1   Illustration for Problem and Solution 7-1.

- Tanks that hold methane gas require periodic inspection and certification by licensed and qualified personnel. This can be inconvenient and costly.

PROBLEM 7-1

Draw a flowchart that outlines a process by which biogas from animal and plant waste (composting) can be produced, distributed, and used.

SOLUTION 7-1

Figure 7-1 illustrates a process by which this can be done on a local or regional scale.

Propane for Propulsion

Propane is also known as *liquefied petroleum gas* (LPG). It, like methane, is a byproduct of petroleum refining, and is known for its use in home heating and cooking. It is also extensively used to power medium-sized electric generators, such as those used by people who own recreational vehicles (RVs) or who have homes operating off the conventional utility grid. As with methane, propane was used to run vehicles before gasoline was available. Today, hundreds of thousands of propane-powered cars and trucks are operational on the roads and highways of the United States.

HOW IT WORKS

Propane and methane work in similar ways when used for propulsion. The main difference is the fact that propane is a liquid in storage, while methane is a gas. Propane becomes gaseous when it is released from the tank. Propane typically contains approximately 8.4×10^4 Btu/gal of energy content.

Engineers sometimes speak of the *gasoline-gallon equivalent* (GGE) of an alternative fuel, which is the ratio of the number of British thermal units (Btu) available in one U.S. gallon (1 gal) of gasoline to the number of British thermal units available in 1 gal of the alternative substance in question. The GGE of propane (let's call it G_p) can be determined as follows:

$$G_p = (1.25 \times 10^5) / (8.4 \times 10^4)$$

$$= 1.5$$

One U.S. gallon of gasoline provides approximately 1.25×10^5 Btu, which is 1.5 times the energy (when burned outright) as 1 gal of liquid propane. The real-world value can vary somewhat from this, depending on the grades of fuel and on their purity. The economic ramifications of any GGE figure depend on the price of the alternative fuel, the price of gasoline, the efficiency of a gasoline-powered engine, and the efficiency of an equivalent alternative-fuel-powered engine.

The GGE figure can be calculated using units other than the gallon and the Btu, as long as the units are the same for both fuels being compared. For example, we can use liters and joules to calculate the GGE of propane, and as long as these units are used to define the energy per unit volume of both propane and gasoline, the resulting GGE figure will still be 1.5. This is because the GGE is a *dimensionless quantity*: a ratio of two parameters that are both expressed in identical units.

The power, cruising speed, and acceleration characteristics of a propane-powered vehicle are similar to those for methane-powered vehicles. It takes about the same amount of time to fill the tank of a propane-powered vehicle as it does to fill a tank of the same size in a conventional fossil-fuel vehicle. The farming of crops for ethanol production takes up CO_2, a known greenhouse gas, from the atmosphere. Most of the light-duty propane-powered vehicles in use today have been converted from gasoline or diesel power. However, a few vendors offer new propane-powered vehicles. There are thousands of propane refueling stations in the United States, but they are not as easy to find as regular gasoline or diesel refueling stations.

ADVANTAGES OF PROPANE FOR PROPULSION

- The emissions from a properly operating propane-powered engine are lower than those from a properly operating gasoline- or diesel-powered engine of the same size. This includes ozone-forming compounds, as well as toxic chemicals such as benzene, formaldehyde, and acetaldehyde.

- At the time of this writing, the cost of propane, considering its GGE, was lower than the cost of gasoline in many locations. That translates into a lower fuel cost overall for propane-powered vehicles as compared with fossil-fuel vehicles.

- Although an extensive infrastructure for propane refueling does not yet exist, the cost of developing one would be lower than the cost of developing a similar infrastructure for certain other alternative energy technologies such as hydrogen or fuel cells.

- Most of the propane used in the United States comes from domestic sources. Thus, the supply and price is less likely to be affected by events overseas.

LIMITATIONS OF PROPANE FOR PROPULSION

- Special training is required to safely refuel, operate, and maintain propane-powered vehicles. Although propane is safer than fuels that are in gaseous form when stored, propane fuel becomes a gas when it enters the atmosphere. This increases the risk of explosion compared with conventional fuels.

- The maximum operating range of a dedicated propane-powered vehicle is only about 67 percent that of the maximum operating range of a gasoline- or diesel-powered vehicle with the same size tank. (This figure comes from the ratio $1/G_p = 1/1.5 = 0.67 = 67\%$.)

- Fuel tanks for propane must be more rugged than those used for gasoline or diesel fuel. This translates into greater vehicle mass, which has an adverse effect on fuel mileage and acceleration characteristics. The greater mass of a propane-powered vehicle also increases the required braking distance.

- Tanks that hold propane require periodic inspection and certification by licensed and qualified personnel. This can be inconvenient and costly.

PROBLEM 7-2
Are hybrid vehicles available that can run on either conventional fuel or propane? Can a gasoline- or diesel-powered vehicle be converted to run on propane?

SOLUTION 7-2
The answer to both of these questions is "Yes." A propane-powered vehicle can be equipped with a hybrid fueling system that will allow it to run on either propane or conventional fuel. It is also possible to equip a gasoline- or diesel-powered vehicle with a hybrid or propane-only fueling system.

Ethanol for Propulsion

Ethanol is produced by fermentation of grains or other plant matter. For this reason it is considered a biofuel, along with biogas and biodiesel. The ethanol production process resembles the way in which liquors such as vodka are made. As a fuel, ethanol is as old as methane and propane. Henry Ford favored the burning of ethanol in his Model T and other early cars. Ironically, he called ethanol "the fuel of the future"! In its pure form, ethanol is a flammable, volatile, clear liquid.

HOW IT IS MADE

In the United States, corn is the most popular crop for producing ethanol. In Brazil, sugar cane is used. In Canada, wheat has been used as the source plant. Theoretically, *cellulose* from any plant matter, terrestrial or aquatic (even trees or seaweed), can be used to obtain ethanol.

The farming of crops for ethanol production takes up CO_2, a known greenhouse gas, from the atmosphere. That same CO_2 is returned to the atmosphere when the ethanol is burned. Therefore, ethanol use is CO_2 *neutral* when it is produced from plants specifically grown for that purpose. However, if plants are cut down to produce ethanol and the demised biomass is not replaced, the use of the resulting fuel causes a net increase in atmospheric CO_2.

Figure 7-2 is a simplified flowchart showing the *dry mill process*, a common method of obtaining ethanol from corn.

HOW IT WORKS

Most modern vehicles can burn gasoline with up to 10 percent ethanol added, and there are no adverse effects. Blends that contain 10 percent or less of ethanol are

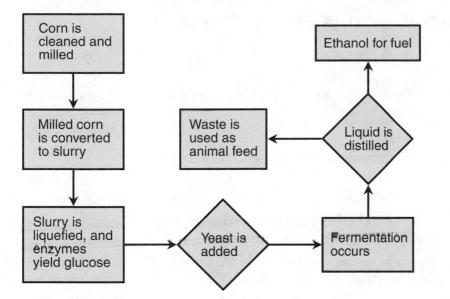

Figure 7-2   The dry mill process is used to derive ethanol from corn, and yields animal feed as a byproduct.

sometimes called *E10* or *gasohol*. If there is more than 10 percent ethanol mixed in with gasoline, vehicle engines require modification in order to burn the fuel efficiently. The percentage of ethanol in some blends is 85 percent, in which case the fuel is known as *E85*.

So-called *flex-fuel vehicles* can burn a variety of ethanol/gasoline mixtures. A sensor, connected to the computer that controls the engine, detects the percentage of ethanol in the fuel tank and adjusts the engine accordingly. The mileage per gallon with ethanol blends is a few percent lower than the mileage with pure gasoline, because ethanol has a lower energy density than gasoline. The extent of this effect depends on the proportion of fuels in the blend; the more ethanol, the worse the mileage. But because added ethanol increases the octane rating of gasoline, performance often improves with gasohol or E85.

During the first years of the 21st century when the price of gasoline in the United States soared, interest grew in E85. Most cars at that time were not equipped to burn E85, and its distribution was limited. Most outlets appeared in the Upper Midwest, where corn is grown in abundance. By the time you read this, E85 should be more popular, its availability should be more widespread, and more cars and trucks should be capable of running on it. Nevertheless, motorists should always check manufacturer's specifications before attempting to burn E85 in any engine.

ADVANTAGES OF ETHANOL FOR PROPULSION

- Ethanol can be used as an extender and octane enhancer in conventional gasoline.
- Ethanol can be used as a primary fuel (in E85), thus reducing dependence on petroleum products.
- Ethanol can reduce emissions of deadly CO gas.
- Ethanol production and use does not contribute to global CO_2 when responsibly done.
- Ethanol can help prevent gas-line freeze in extremely cold weather.
- In some locations, gasohol and E85 are cheaper per gallon than conventional gasoline.
- The widespread production and use of ethanol as a fuel benefits farmers by increasing the demand for their products.
- The plant matter used to produce ethanol is a renewable resource.
- Ethanol is not as flammable as gasoline, and is therefore less likely to cause accidental explosions.

LIMITATIONS OF ETHANOL FOR PROPULSION

- Fueling stations for gasoline containing ethanol, and especially for E85, are not (at the time of this writing) as abundant as fueling stations that offer only conventional gasoline.
- In some locations, gasohol and E85 are more expensive per gallon than conventional gasoline.
- Some vehicles can be damaged by inappropriate use of ethanol fuel. This is especially true if E85 is inadvertently pumped into a vehicle designed to burn conventional gasoline.
- Because some, if not most, ethanol is produced from plant matter that could otherwise be consumed as human food, some people argue that the widespread use of ethanol as fuel will contribute indirectly to world hunger.

PROBLEM 7-3
Can the ethanol from the fuel refining process be used to mix intoxicating drinks?

SOLUTION 7-3
Don't try to use fuel ethanol to spike drinks! The law requires that a small amount of gasoline or other objectionable additive be mixed in with ethanol when it is produced for use as fuel. If you consume this "pure ethanol," it will make you sick. Get your drinking alcohol at places where liquors are sold for human consumption, not for use in internal-combustion engines!

Biodiesel for Propulsion

Biodiesel is a combustible, rather viscous liquid consisting of alkyl esters of fatty acids derived from vegetable oil or cooking grease. This fuel is designed to be used in compression-ignition engines similar or identical to those that burn petroleum diesel.

HOW IT IS MADE

Soybeans are the cheapest and most abundant vegetable source of oil that can be refined to obtain biodiesel. The extracted oil is processed to remove all traces of water, dirt, and other contaminants. Free fatty acids are also removed. A combination of methyl alcohol and a catalyst, usually sodium hydroxide or potassium hydroxide, breaks the oil molecules apart in a chemical reaction known as *esterification*. The resulting compounds, called *esters*, are then refined into usable biodiesel. Figure 7-3 is a simplified flowchart showing this process.

Used-up cooking oil and animal fats, which would otherwise be thrown away, can also be refined to produce biodiesel. The process is similar to the way biodiesel is derived from soybean oil, except there is an additional step involved (Figure 7-4). Methyl alcohol and sulfur are used in a process called *dilute acid esterification* to obtain a substance resembling fresh vegetable oil, which is then processed in the same way as soybean oil to obtain the final product.

Biodiesel is commonly blended with petroleum diesel. The percentage of biodiesel in the blend is written following an uppercase letter *B* to denote the proportion. For example, a mixture of 20 percent biodiesel and 80 percent petroleum diesel is called B20, a mixture of equal parts biodiesel and petroleum diesel is B50, and pure (or *neat*) biodiesel is B100.

HOW IT WORKS

Nearly all conventional diesel engines can burn blends from pure petroleum diesel up to B20 without modification. In most diesel engines built since 1994,

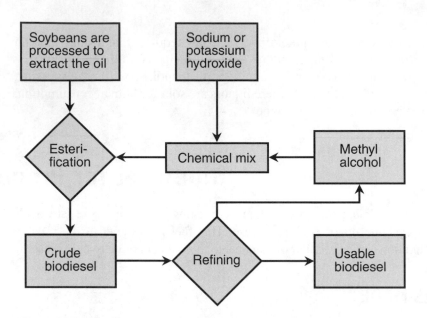

Figure 7-3   Production of biodiesel from soybeans.

blends from B20 to B100 can be used with minor modification, but transportation and storage of these fuels require special attention. It is recommended that owners of all diesel engines review their warranty statements before attempting to burn any substitute for pure petroleum diesel. Some manufacturers will void the warranty if an engine is run on fuel higher than a certain percentage of biodiesel.

The combustion of biodiesel produces less CO_2 gas than the combustion of petroleum diesel. In addition, biodiesel contains less sulfur, particularly when derived from vegetable sources such as soybeans. The result is reduced emissions of compounds such as sulfur dioxide (SO_2), which is known to contribute to environmental damage in the form of acid precipitation. Most other emissions are reduced as well, including deadly CO gas and particulate matter. However, nitrous oxide emissions are increased with biodiesel as compared with petroleum diesel.

Biodiesel, when burned outright, yields approximately 1.17×10^5 Btu/gal of energy. The GGE of biodiesel (let's call it G_{bd}) can therefore be defined as follows:

$$G_{bd} = (1.25 \times 10^5) / (1.17 \times 10^5)$$

$$= 1.07$$

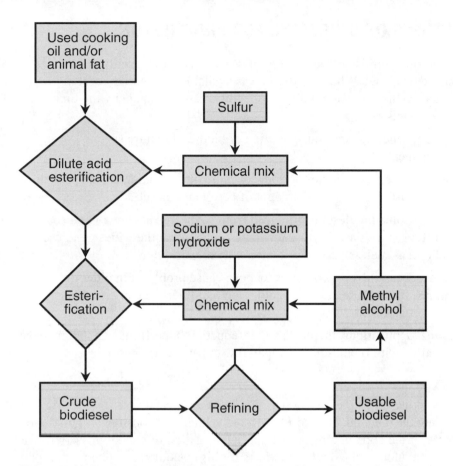

Figure 7-4 Production of biodiesel from used cooking oil and animal fat.

ADVANTAGES OF BIODIESEL FOR PROPULSION

- Biodiesel combustion produces less emissions (with the exception of nitrous oxides) than combustion of an equal amount of petroleum diesel.

- Biodiesel comes from renewable resources. The supply can be literally grown! But petroleum diesel comes from resources that are not renewable.

- Defunct cooking oils and grease can be used to produce biodiesel.

- The widespread use of biodiesel can reduce dependency on imported oil.

- Pure biodiesel is not toxic if spilled, because it is biodegradable. Petroleum diesel can cause environmental damage if spilled in large amounts.

- Biodiesel is safer than petroleum diesel. It is less combustible when stored or transported.

LIMITATIONS OF BIODIESEL FOR PROPULSION

- At the time of this writing, biodiesel was not as widely available as petroleum diesel. Where biodiesel was available, it generally cost more per gallon than petroleum diesel. (Conditions may change by the time you read this.)

- Storage, handling, and transportation of biodiesel require special management.

- The combustion of biodiesel produces more nitrous oxide emissions than the combustion of an equal amount of petroleum diesel.

- Because some biodiesel is produced from soybeans that are a good source of protein as well as oils, there is some concern that the widespread use of biodiesel as fuel will contribute indirectly to world hunger.

- Biodiesel has solvent properties that can cause problems in older diesel engines. It can loosen deposits, clogging fuel filters. It can also damage rubber components.

- In most applications, there is a slight reduction in performance and mileage per gallon with biodiesel as compared with petroleum diesel.

PROBLEM 7-4

Can used cooking oil or grease, such as that from a deep-fat fryer or a frying pan, be poured into the fuel tank of a diesel-powered vehicle and consumed as biodiesel? For example, if the cook at a restaurant has a half gallon of bacon grease left over from the preparation of the day's breakfasts for his customers, can he pour the hot grease straight into the tank of his diesel truck and expect the grease to function as biodiesel?

SOLUTION 7-4

Absolutely not! Demised cooking oil or grease must be processed as shown in Figure 7-4 before it can be used as biodiesel. This should be obvious in the case of bacon grease, which solidifies near room temperature. But it is true even of fats that remain liquid at relatively low temperatures, such as corn oil, canola oil, or even soybean oil.

Quiz

This is an "open book" quiz. You may refer to the text in this chapter. A good score is eight correct. Answers are in the back of the book.

1. Biogas is another name for

    a. gasoline produced from soybeans.

    b. propane produced from fermentation of corn.

    c. methane produced from composting.

    d. biodiesel produced from animal fat.

2. Suppose the GGE of an alternative liquid fuel is exactly equal to 2. That means that if a specific volume of gasoline yields x Btu of energy when burned outright, the alternative fuel, when burned outright, yields

    a. $4x$ Btu.

    b. $2x$ Btu.

    c. $0.5x$ Btu.

    d. $0.25x$ Btu.

3. A fuel called B70 would consist of

    a. 70 percent ethanol.

    b. 70 percent gasoline.

    c. 30 percent biodiesel.

    d. None of the above

4. Fill in the blank to make the following sentence true: "Petroleum diesel engines built prior to 1994 should be able to burn _____ without modification."

    a. B10

    b. B40

    c. B70

    d. B100

5. In Brazil, ethanol is obtained largely from

    a. sugar cane.

    b. soybeans.

    c. corn.

    d. coconuts.

6. With respect to which of the following emission types is petroleum diesel a cleaner-burning fuel than biodiesel?

  a. Sulfur dioxide

  b. Carbon dioxide

  c. Carbon monoxide

  d. Nitrous oxides

7. Which of the following statements about biodiesel is false?

  a. Biodiesel combustion produces less sulfur dioxide than petroleum diesel combustion.

  b. Biodiesel can be produced from discarded cooking grease.

  c. Biodiesel comes from renewable resources.

  d. Biodiesel is a byproduct of the production of propane.

8. Which of the following fuels is normally a gas when stored?

  a. Propane

  b. Biodiesel

  c. Methane

  d. None of the above

9. Ethanol/gasoline mixtures of various proportions can be burned in

  a. diesel engines.

  b. flex-fuel vehicles.

  c. propane engines.

  d. All of the above

10. A liter of propane burned outright provides

  a. 1.5 times as much energy as a liter of gasoline burned outright.

  b. about the same amount of energy as a liter of gasoline burned outright.

  c. about two-thirds as much energy as a liter of gasoline burned outright.

  d. about half as much energy as a liter of gasoline burned outright.

CHAPTER 8

Propulsion with Electricity, Hydrogen, and Fuel Cells

The demand for petroleum and its derivatives will increase in the coming years. China and India will emerge as major economic powers. Many experts believe that the supply of so-called fossil fuels will not be adequate to meet the skyrocketing demand. This is one of the strongest arguments for development of "true alternatives" to fossil fuels for transportation. Electricity is a promising solution. One of the oldest alternative propulsion systems makes use of battery-powered direct-current (DC) motors. More sophisticated systems derive the electricity from *fuel cells*.

Electric Vehicles

The basic concept behind the *electric vehicle* (EV) is straightforward. A motor is powered by a source of DC. This source can be a massive, rechargeable battery that is carried along with the vehicle.

HOW THEY WORK

In an EV, the motor is connected to the wheels by a *drive system* similar to the *transmission* found in an ordinary car or truck. The speed of the vehicle depends on the speed at which the motor runs. The motor speed is controlled by an accelerator pedal (just like the one in any other car or truck), or by a handlebar lever in the case of an electric motorbike or all-terrain vehicle (ATV). The speed of the vehicle also depends on the gear ratio between the motor and the wheels. The motor gets its power from a rechargeable battery. The two most common types are the *lead-acid battery* and the *nickel-based battery*.

Most electric cars can travel about 100 km (60 mi) on a full charge before recharging is required. Improved designs may double this figure. However, the actual distance depends on the nature of the terrain, whether the vehicle is going into the wind or against it, and the temperature of the outdoor air. Maximum attainable speeds are on the order of 130 km/h (approximately 80 mi/h) under ideal operating conditions: flat terrain, no wind, and a temperature of about 20°C (70°F).

LEAD-ACID CELLS AND BATTERIES

Figure 8-1 is a functional diagram of a *lead-acid cell*. A plate of lead serves as the negative electrode, and a plate of lead dioxide serves as the positive electrode. Both electrodes are immersed in a sulfuric-acid solution called the *electrolyte*. The result is that a *potential difference*, more often called a *voltage*, develops between the electrodes. This voltage can drive current through a load. The *maximum deliverable current* depends on the mass and volume of the cell. In a lead-acid battery made from lead-acid cells connected in series (negative-to-positive), the voltage depends on the number of cells.

If a lead-acid battery is connected to a load for a long time, the current gradually decreases, and the electrodes become coated. The nature of the acid changes, too. Eventually, all the chemical energy contained in the acid is converted into electrical energy. Then the current drops to zero, and a potential difference no longer exists between the electrodes in the cells. However, if a current is driven through the battery by connecting an external source of DC to the battery terminals for a period of time (negative-to-negative and positive-to-positive), the electrochemical energy in the battery is replenished, and the battery can be used again. This cycle can be repeated many times over the useful life of a battery.

Lead-acid batteries are used in nearly all motor vehicles to provide the power for the initial startup ignition. They are also found in *uninterruptible power supplies* for computer workstations, in pellet stoves to keep the blowers running in the event

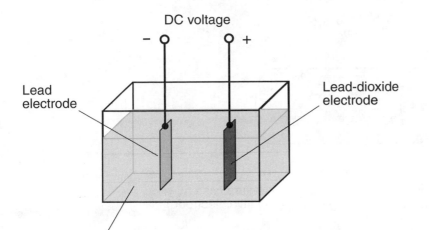

Figure 8-1 Simplified diagram of a lead-acid cell. When multiple cells are combined in series and in parallel, the result is a lead-acid battery.

of a utility power failure, and in *notebook computers* as the main power source during portable operation.

NICKEL-BASED CELLS AND BATTERIES

Nickel-based cells include the *nickel-cadmium* (NICAD or NiCd) type, shown in Figure 8-2A, and the *nickel-metal-hydride* (NiMH) type, shown in Figure 8-2B. The two types are identical except for the composition of the negative electrode. *Nickel-based batteries* are available in packs of cells connected in series, in parallel, or in series-parallel to obtain higher current and voltage ratings than is possible with a single cell. All nickel-based cells are rechargeable, and can be put through hundreds of charge/discharge cycles if they are properly cared for.

Nickel-based cells are found in various sizes and shapes. *Cylindrical cells* look like ordinary dry cells. These are the types diagrammed in Figure 8-2. *Button cells* are small, pill-shaped units commonly used in cameras, watches, memory backup applications, and other places where miniaturization is important. *Flooded cells* are designed for heavy-duty applications and lend themselves to use in EVs. These are the types that best lend themselves to use in EVs. *Spacecraft cells* are made in packages that can withstand the rigors of a deep-space environment.

A nickel-based cell or battery should never be discharged all the way until it "totally dies." This can cause the polarity of one or more of the cells to reverse. Once this happens, the battery is ruined.

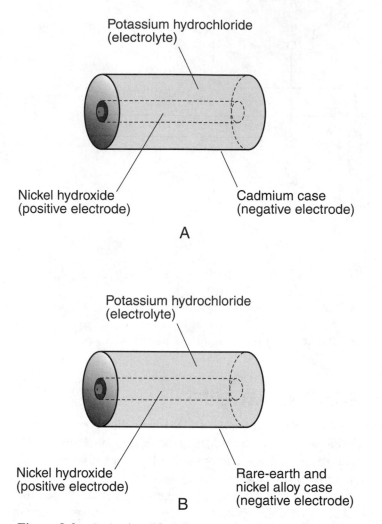

Potassium hydrochloride
(electrolyte)

Nickel hydroxide
(positive electrode)

Cadmium case
(negative electrode)

A

Potassium hydrochloride
(electrolyte)

Nickel hydroxide
(positive electrode)

Rare-earth and
nickel alloy case
(negative electrode)

B

Figure 8-2   At A, simplified diagram of a nickel-cadmium (NICAD) cell. At B, simplified diagram of a nickel-metal-hydride (NiMH) cell.

A discharge anomaly peculiar to NICAD cells and batteries is known as *memory* or *memory drain*. If a NICAD unit is used over and over, and is discharged to the same extent every time, it will lose much of its current-delivering capacity at that point in its discharge cycle. This gives the impression that the battery is no longer any good. Memory problems can usually be solved. Use the cell or battery almost all the way up, and then fully recharge it. Repeat the process several times. The newer NiMH cells and batteries do not suffer as often as older NICADs from memory drain. In addition, NiMH units can store up to 40 percent more energy than NICAD units having the same mass and volume.

STORAGE CAPACITY

Any cell or battery has a certain amount of electrical energy that can be specified in watt-hours (Wh) or kilowatt-hours (kWh). Often it is given in terms of the mathematical *integral* of deliverable current with respect to time, in units of *ampere-hours* (Ah). The energy capacity in watt-hours is the ampere-hour capacity multiplied by the battery voltage.

Consider a battery with a rating of 1000 Ah. This battery can deliver 100 A for 10 h, or 10 A for 100 h, or 1 A for 1000 h. There exist an infinite number of current/time combinations, and almost any of them (except for the extremes) can be put to use in real life. The extreme situations are the *shelf life* and the *maximum deliverable current*. Shelf life is the length of time the battery will remain usable if it is never connected to a load. The maximum deliverable current is the highest current a battery can drive through a load without the voltage dropping significantly because of the battery's internal resistance, and without causing the battery to overheat.

BATTERY CHARGING

Storage batteries can be recharged in a variety of ways. The most common method is the use of an external charging unit connected to a source of conventional utility power such as a wall outlet. An alternative-energy charging system using a wind turbine, water turbine, or solar panel can also be used. Solar panels can be placed on the roof, trunk lid, or hood to provide some supplemental charging when the vehicle is exposed to sunlight. In more advanced vehicles, energy for recharging can be derived from a primary internal combustion engine and an *alternator*, a set of *fuel cells*, or a special braking system that slows the vehicle down by "robbing" kinetic energy from the wheels and using it to charge the battery.

THE POLLUTION QUESTION

At first thought, it is tempting to suppose that EVs do not pollute, but it's not that simple. Lead-acid and NICAD batteries can damage the environment if they are carelessly discarded. Besides that, the energy for charging them has to come from somewhere, and that is usually a polluting electric utility plant.

The sulfuric acid in a lead-acid battery produces fumes including sulfur dioxide, a known pollutant gas. Hydrogen, a flammable gas that can explode if confined and exposed to flame or spark, is also produced. Lead and cadmium are heavy metals, and constitute known environmental toxins. Therefore, special precautions must be taken when discarding old lead-acid or NICAD batteries. Because of pollution concerns, NiMH batteries have replaced NICAD types in many applications. In most practical scenarios, a NICAD battery can be directly replaced with a NiMH battery of the same voltage and current-delivering capacity, and the powered-up device will work satisfactorily.

When the battery in an EV is charged, it consumes a little more energy than the actual energy derived from the battery over the course of the discharge cycle. A 12-V battery rated at 2000 Ah will require a little more than 24,000 Wh, or 24 kWh, of energy to acquire a full charge. That is the equivalent of a portable electric space heater running on the "high" setting for 16 h. This amount of energy will run a compact EV car in the city for approximately 80 km (50 mi) under ideal conditions.

If the number of EVs in use were to increase, there would be a corresponding increase in the demand for electricity from power plants, many of which use methane, oil, or coal to run the generators. Even so, it is generally agreed that if EV usage were to displace conventional vehicle usage kilometer-for-kilometer without any increase in the total number of kilometers driven by the population, overall pollution would diminish. According to some estimates, the overall pollution generated by EVs, kilometer-for-kilometer, is only about 10 percent of the overall pollution generated by fossil-fuel vehicles.

ADVANTAGES OF EVS

- The use of EVs can help industrialized countries to reduce dependency on foreign oil.

- The overall pollution generated by EVs, kilometer-for-kilometer, is a fraction of that generated by vehicles that use combustion engines, even when the pollution from utility plants and battery manufacture is taken into consideration.

- The cost of the energy required to operate an EV, kilometer-for-kilometer, is lower than the cost of the energy required to operate a fossil-fuel vehicle.

- In some locations, tax breaks or rebates are available for people who use alternative vehicles, including EVs.

- With some ingenuity, EVs can be partially or completely charged from sources other than the electric utility.

- The use of an EV can provide the user with a sense of independence.

LIMITATIONS OF EVS

- The maximum operating range for an EV, starting with a fully charged battery, is less than the operating range of a typical fossil-fuel vehicle.

- The battery in an EV can lose some of its ability to hold a charge or deliver sufficient current when the temperature falls far below zero Celsius.

- An EV may not be powerful enough to operate reliably in severe weather, particularly in heavy snow.

- The EV design concept does not lend itself well to heavy-duty applications such as hauling freight or plowing snow.

- Because most EVs employ light construction and are small in size, safety can be a concern.

- In some areas, it may be difficult to get EVs serviced because of a lack of parts or competent technicians.

PROBLEM 8-1

How can the interior of an EV be kept at a comfortable temperature when the outdoor air is extremely cold or extremely hot?

SOLUTION 8-1

Modern EVs employ air-source heat pumps to maintain habitable interior temperature. The interior can be preheated or precooled while the EV is connected to the charging station. Enhanced thermal insulation can also help. But in extreme conditions, particularly when the outdoor air is dangerously cold, maintaining a comfortable interior temperature while the vehicle is in operation reduces the operating range between charges.

Hybrid Electric Vehicles

Electric and conventional energy sources are combined in the *hybrid electric vehicle* (HEV). There are two basic types of HEV, known as the *series design* and the *parallel design*.

SERIES HEVS

In a series HEV, the mechanical power is obtained from an electric motor powered by a rechargeable battery. In this sense, it resembles an EV. But instead of relying on an external source of energy for battery charging, the series HEV carries its charging system on board as a generator driven by a small engine. The engine and generator run continuously at constant speed while the vehicle is in operation. The engine can be powered by gasoline, methane, propane, E85, petroleum diesel, biodiesel, or hydrogen.

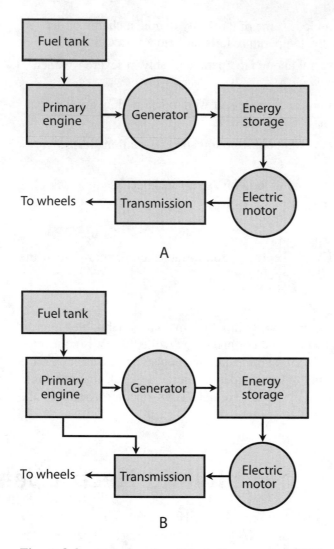

A

B

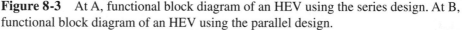

Figure 8-3 At A, functional block diagram of an HEV using the series design. At B, functional block diagram of an HEV using the parallel design.

Figure 8-3A is a block diagram that shows the general configuration of a series HEV. The generator includes a charging unit in the form of an alternating-current-to-direct-current (AC/DC) converter. The energy storage medium is a large rechargeable battery, of the same type used in an EV. An optional external charger can be used when the vehicle is not running to keep the battery "topped off." The electric motor is connected to a transmission that resembles the drive system in an EV.

The distinguishing feature of the series HEV is the fact that it always runs from the electric motor, and never from a fuel-powered combustion engine. For this reason, the speed and acceleration capabilities of the series HEV are limited. Electric motors, in general, do not provide the short-burst high power that can be obtained with combustion engines. The primary asset of the series design is the fact that the combustion engine is small, and this translates into low fuel usage and minimal polluting emissions.

PARALLEL HEVS

The outstanding feature of the parallel HEV is the fact that both the primary engine and the electric motor contribute directly to propulsion. They operate in tandem (parallel), with the burden automatically shifting from one to the other, depending on driving conditions from moment to moment. The primary engine is usually powered by conventional gasoline or gasohol, although alternative fuels such as methane or propane can be used.

Figure 8-3B shows the interconnection of components in a parallel HEV. As in the series design, the generator includes an AC/DC converter to charge the battery, and an optional external charger can be used when the vehicle is parked. The electric motor and primary engine are both connected to the transmission. At low speed, or when relatively little power is required to propel the vehicle, the electric motor is used. At high speed, or when a lot of power is needed (when going up a steep hill or when passing another vehicle, for example), the internal combustion engine is used. A microcomputer determines which engine is required in greater proportion, and shifts the burden based on the amount of *back pressure* in the drive system.

REGENERATIVE BRAKING

In EVs and HEVs, the battery can derive extra charging energy by a scheme called *regenerative braking* when the vehicle is decelerating or traveling down a steep grade. A generator and an AC/DC converter, connected into the drive chain, harnesses some of the energy that would otherwise be wasted heating the brake drums. When the wheels are forced to turn against the resistance of a generator connected to a load, the effect is to *decelerate* the vehicle (slow it down), just as brakes do.

In an EV or HEV equipped with regenerative braking, a microcomputer determines when the situation is such that power should be taken from the wheels rather than delivered to them. The wheels then supply the mechanical torque for the battery-charging generator, reducing the amount of energy that would otherwise be necessary to charge the battery from the primary engine or the external charger. This also reduces wear on the brakes, extending their life.

ADVANTAGES OF HEVS

- The use of HEVs can help industrialized countries to reduce dependency on foreign oil.

- The overall pollution generated by HEVs, kilometer-for-kilometer, is less than that generated by vehicles that use combustion engines, even when the pollution from utility plants and battery manufacture is taken into consideration.

- The cost of the energy required to operate an HEV, kilometer-for-kilometer, is less than the cost of the energy required to operate a fossil-fuel vehicle, after the up-front investment has been made.

- In some locations, tax breaks or rebates are available for people who use alternative vehicles, including EVs.

- An HEV can operate from gasoline or alone if necessary, and may also get its owner through periods when the supply of gasoline is cut off.

- The parallel HEV concept can be used in the design of heavy-duty vehicles such as buses and large trucks. An increasing number of such vehicles are adopting this technology.

LIMITATIONS OF HEVS

- A typical HEV costs more, in terms of the purchase price, than a conventional vehicle of the same size. (However, the gap is narrowing.)

- Because of the complexity of their design, it may be difficult to find competent technicians who can service HEVs. This is particularly true in rural areas, or in other regions where HEVs are rarely used.

- Many of the components in an HEV are specialized or proprietary, and are not as easily available in remote locations as are the parts for old-fashioned internal-combustion-engine vehicles.

- In extremely cold weather, the battery in an HEV may not operate properly.

PROBLEM 8-2
Is there a way to store electrical energy derived from braking or downhill coasting, without involving the battery, if an intense burst of power is called for a short while later?

SOLUTION 8-2
Yes. A device called an *ultracapacitor* can do this. It operates on the principle of separation of charge, in the same way as the capacitors in DC power supplies

function in electronic equipment. An ultracapacitor can store far more electrical charge than an ordinary capacitor, and can deliver a large current for a short time to provide an intense burst of mechanical power from the electric motor. These devices are employed in some sophisticated HEVs.

Hydrogen-Fueled Vehicles

When burned in the presence of pure oxygen, hydrogen, the lightest and most abundant element in the universe, liberates only energy and water vapor. When burned in the atmosphere, which is approximately 71 percent nitrogen, some nitrous oxide gas is produced as well. Conventional fossil-fuel vehicles can be adapted to run on hydrogen, just as they can be adapted to run on methane or propane. The term *hydrogen-fueled vehicle* is sometimes abbreviated as HFV.

GETTING THE HYDROGEN

Hydrogen must be separated out from other compounds in nature. It does not naturally exist in its free state. The most common way to obtain hydrogen is from methane by a process known as *steam reforming*. In steam reforming, methane reacts with steam in the presence of elemental nickel, which acts as a catalyst. In addition to the hydrogen, CO gas is released as a byproduct.

Another way to obtain hydrogen is by *electrolysis*. A molecule of pure water contains two atoms of hydrogen and one atom of oxygen chemically bound together. This bond can be split if an electric current passes through water. Electrolysis requires the addition of an *electrolyte*, such as salt, sodium bicarbonate, or sulfuric acid, to the water to enhance its ability to conduct electricity. When a DC voltage is applied between two electrodes immersed in this solution, hydrogen bubbles appear at the negative electrode, while oxygen bubbles appear at the positive electrode (see Figure 8-4). The gas from these bubbles is collected and stored. No dangerous byproducts are generated. It is necessary to use DC, not AC such as that from household utility outlets. If AC is used, hydrogen and oxygen both appear at both electrodes, and it is impossible to separate them.

Electrolysis is an expensive and inefficient process. It takes half again as much electrical energy to produce hydrogen by electrolysis as the resulting hydrogen yields if burned outright. In order to produce hydrogen on a large scale by means of electrolysis, dedicated solar or wind power plants have been suggested. It doesn't take much water to produce a lot of hydrogen, and saline or "alkali" lakes exist in various places that already contain the necessary electrolyte substances.

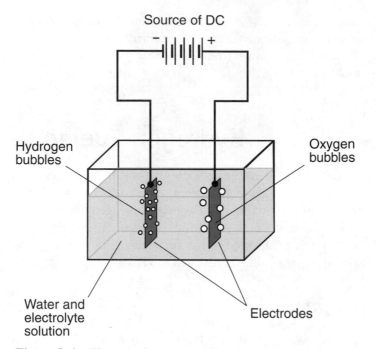

Figure 8-4   Electrolysis separates liquid water into hydrogen and oxygen gases.

Alternative methods of electrolysis can break water molecules apart to obtain hydrogen. These methods involve the use of certain chemicals and/or heat. Another way to get hydrogen is to break down coal or biomatter, substances largely consisting of *carbohydrates*, into their elemental components.

STORING THE HYDROGEN

Onboard fuel storage is another technical problem that will have to be solved before hydrogen becomes practical for use as a motor-vehicle fuel. Hydrogen can be compressed in metal tanks, just as can other gases. But to provide enough fuel to propel a hydrogen vehicle for a reasonable distance between refueling stops, the tank must be so large and massive that it impairs vehicle efficiency. In addition to this, compressed-gas tanks can be dangerous. In the event of a collision that causes the neck of the tank to break, the resulting release of pressure can turn the tank into a deadly missile. One of the main objectives of research-and-development efforts, therefore, is to find a safe and effective way to store hydrogen at high pressure in a tank of reasonable size and mass.

Other fuel storage methods are under consideration. One technology involves the use of *metal hydrides* (compounds of metal and hydrogen), from which the hydrogen can be liberated at a controlled rate under certain conditions. Another technology makes use of a phenomenon called *gas-on-solid adsorption*. In this process, hydrogen is stored in complexes of extremely small carbon structures called *nanotubes*. Hydrogen gas can condense inside these structures at density levels comparable with the best storage tanks, but without the dangers of high pressure associated with tanks. Controlled liberation is possible, and the resulting hydrogen gas can be directly utilized as a fuel source. Both of these technologies, as of this writing, are in the research-and-development phase.

ADVANTAGES OF HFVS

- Hydrogen, when burned with pure oxygen, is totally nonpolluting.

- Hydrogen engines are more efficient than methane or propane engines in converting the energy from the fuel into mechanical energy. In fact, hydrogen engines are comparable with gasoline engines in this respect.

- The widespread production and use of hydrogen may mitigate problems, such as variable prices and supply interruptions, that occur with conventional fuels.

- Increased production of hydrogen for use in vehicles would, as a side benefit, increase the hydrogen supply available for heating homes and businesses.

- In many countries, including the United States, distribution pipelines for methane gas already exist. Some of these could be adapted to hydrogen for use in a network of refueling stations.

- Hydrogen can be produced in small-scale and local facilities, as well as at centralized facilities. This could enhance the security of the civilized world by distributing energy resources and assets.

LIMITATIONS OF HFVS

- In its combustible form, and at ordinary room temperature and pressure, hydrogen is a gas. This presents storage, handling, and transportation problems. In particular, designing a safe fuel tank for a hydrogen-powered vehicle is a challenge.

- Hydrogen was not widely available at vehicle refueling stations in most countries at the time of this writing. This included the United States.

- Tanks that hold hydrogen gas require periodic inspection and certification. This must be done by licensed and qualified personnel.

- At the time of this writing, hydrogen was a comparatively expensive fuel, mainly because of the cost of the processes involved in separating it out from naturally occurring compounds such as methane or water.

PROBLEM 8-3

How far can an HFV be driven on the amount of hydrogen derived, by electrolysis, from a liter of water?

SOLUTION 8-3

An efficient HFV gets about the same mileage from the hydrogen contained in a liter of water as a comparable fossil-fuel vehicle gets from a liter of gasoline.

Fuel-Cell Vehicles

Hydrogen, methane, propane, and other fuels can be burned directly to obtain propulsion, as we have seen. However, these fuels—and, in fact, any fuel—can be utilized indirectly in a specialized form of EV known as a *fuel-cell vehicle* (FCV).

WHAT IS A FUEL CELL?

In the late part of the 20th century, a new type of electrochemical power device emerged that holds promise as an alternative energy source: the *fuel cell*. The most talked-about fuel cell during the early years of research and development became known as the *hydrogen fuel cell*. As its name implies, it derives electricity from hydrogen. The hydrogen combines with oxygen (that is, it *oxidizes*) to form energy and water, along with a small amount of nitrous oxide if air is used as the oxidizer. When a hydrogen fuel cell "runs out of juice," all that is needed is a new supply of hydrogen.

Instead of combusting, the hydrogen in a fuel cell oxidizes in a more controlled fashion, and at a much lower temperature. There are several schemes for making this happen. The *proton exchange membrane* (PEM) *fuel cell* is one of the most widely used. A PEM hydrogen fuel cell generates approximately 0.7 V DC, or a little less than half the voltage of a typical electrochemical dry cell. In order to obtain higher voltages, individual cells are connected in series. When this is done, the voltages of the individual cells add up. For example, to obtain 14 V DC, it is necessary to connect 20 hydrogen fuel cells in series. A series-connected set of fuel cells is technically a battery, but the more often-used term is *stack*. Increased current-delivering capacity can be obtained by connecting cells or stacks in parallel. When this is done, the current-delivering capacities of the individual cells or stacks add up. For example, if five stacks, each rated at 14 V DC and capable of delivering up to 10 A, are connected in parallel, the resulting combination will provide 14 V DC at up to 50 A.

Fuel-cell stacks are available in various sizes. A stack about the size and weight of an airline suitcase filled with books can power a subcompact electric car. Smaller cells, called *micro fuel cells*, can provide electricity to run devices that have historically operated from conventional cells and batteries. These include portable radios, lanterns, and notebook computers.

Hydrogen is not the only substance that can be used to make a fuel cell. Almost anything that will combine with oxygen to form energy has been considered. *Methanol*, a form of alcohol, has the advantage of being easier to transport and store than hydrogen, because it exists as a liquid at room temperature. Propane is another chemical that has been used for powering fuel cells. This is the substance that is stored in liquid form in tanks for barbecue grills and some rural home heating systems. Methane has been used as well. Some scientists and engineers object to the use of propane and methane because they are fuels on which society has developed dependence that purists would like to get away from.

HOW FCVS WORK

An FCV is essentially an EV that uses a fuel cell in place of, or in addition to, a storage battery. Figure 8-5 is a functional block diagram of a hydrogen FCV. The electric motor is connected to the wheels through a drive system similar to the transmissions found in fossil-fuel vehicles. The speed of the vehicle, which is controlled by an accelerator pedal, depends on the speed at which the motor runs, and also on the gear ratio between the motor and the wheels.

The motor gets its power from the electricity provided by the fuel cell, and also, in some designs, from a rechargeable storage battery. If a battery is used, it can derive its charge from the fuel cell, and also from a regenerative braking system of the type used in advanced EVs and HEVs. The use of a storage battery has a special advantage. If the fuel tank is allowed to go empty, the battery can provide some extra driving range.

A typical FCV can convert about 50 percent of the energy contained in the hydrogen gas into usable electrical energy. The remainder of the energy is converted into heat; this must be dissipated by a cooling system that resembles the radiator in a fossil-fuel vehicle. However, because hydrogen burns cleaner than conventional fossil fuels, it is reasonable to expect that a well-designed hydrogen FCV will operate more efficiently than a traditional fossil-fuel vehicle, although perhaps at a lower level of efficiency than a well-designed HFV that burns hydrogen directly to obtain propulsion.

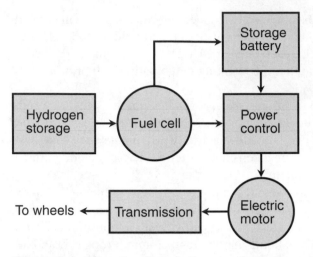

Figure 8-5   Functional block diagram of a hydrogen FCV.

ADVANTAGES OF FCVS

- The use of FCVs can help industrialized countries reduce dependency on foreign oil.

- Increased production of hydrogen for use in FCVs would, as a side benefit, increase the hydrogen supply available for heating homes and businesses.

- Existing methane pipelines could be adapted for use in a network of FCV refueling stations, using either methane or hydrogen as the fuel.

- The overall pollution generated by hydrogen FCVs, kilometer-for-kilometer, is a fraction of that generated by fossil-fuel vehicles, even when the pollution from hydrogen production and battery manufacture is taken into consideration.

- Hydrogen can be produced in small-scale and local facilities.

- In some locations, tax breaks or rebates are available for people who use alternative vehicles, including FCVs.

LIMITATIONS OF FCVS

- The maximum operating range for a hydrogen FCV, starting with a full fuel tank, is less than the operating range of a fossil-fuel vehicle. If other types of fuel cells (methane or propane, for example) are used, the operating ranges of FCVs and fossil-fuel vehicles are comparable.

- In some areas, it may be difficult to get FCVs serviced because of a lack of parts or competent technicians.

- The onboard storage of fuel for hydrogen FCVs presents a major technological obstacle to the widespread deployment of these vehicles.

- Hydrogen FCVs are relatively expensive to operate, largely because of the cost of the processes involved in separating hydrogen from naturally occurring compounds.

PROBLEM 8-4

If hydrogen gas can be burned directly as fuel for a combustion engine in an HFV, what is the point of converting the energy from the hydrogen into electrical energy to run a motor in a hydrogen FCV? Isn't that an unnecessary step that introduces inefficiency into an otherwise workable system?

SOLUTION 8-4

There are two reasons why a hydrogen FCV may be preferred over an HFV. First, the fuel cell operates at a lower temperature than a hydrogen combustion engine. This makes the process inherently safer. In addition, a hydrogen FCV can employ an onboard storage battery, from which the electric motor can directly obtain energy if necessary. The battery also allows for storage of energy derived from regenerative braking. Thus, the hydrogen FCV embodies some of the assets of EVs, HEVs, and HFVs all together.

Quiz

This is an "open book" quiz. You may refer to the text in this chapter. A good score is eight correct. Answers are in the back of the book.

1. Which of the following statements about hydrogen is false?

    a. When burned with pure oxygen, it produces only energy and water.

    b. It is the lightest chemical element in the universe.

    c. It is readily available in free natural form along with oil and natural gas.

    d. It is the most abundant chemical element in the universe.

2. How many PEM hydrogen fuel cells must be connected in series in order to obtain a stack that produces approximately 50 V DC?

   a. 72

   b. 36

   c. 18

   d. You can't get 50 V DC by connecting PEM hydrogen fuelcells in series.

3. Consider the apparatus shown in Figure 8-6. Suppose it is intended to be used for electrolysis. What, if anything, is wrong with it?

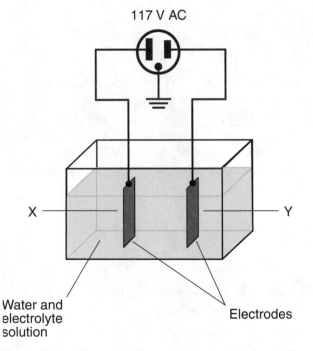

Figure 8-6 Illustration for quiz questions 3, 4, and 5.

   a. Nothing.

   b. The voltage is too high.

   c. The polarity of the voltage is incorrect.

   d. The voltage should be DC, not AC.

4. If the device of Figure 8-6 is used exactly as shown, what will appear at electrode X?

    a. Nothing.

    b. Hydrogen bubbles.

    c. Oxygen bubbles.

    d. Hydrogen and oxygen bubbles.

5. If the device of Figure 8-6 is used exactly as shown, what will appear at electrode Y?

    a. Nothing.

    b. Hydrogen bubbles.

    c. Oxygen bubbles.

    d. Hydrogen and oxygen bubbles.

6. A lead-acid battery in an HEV cannot be recharged by means of

    a. an external charging unit.

    b. electrolysis.

    c. solar panels on the roof.

    d. regenerative braking.

7. Which of the following is an advantage of a NiMH battery over a NICAD battery?

    a. A NiMH battery can produce AC as well as DC.

    b. A NiMH battery has greater energy storage capacity per unit volume.

    c. A NiMH battery exhibits a greater degree of memory drain.

    d. None of the above; NICAD batteries are superior to NiMH batteries.

8. Which of the following represents a significant disadvantage of hydrogen as a fuel for direct combustion with pure oxygen in a motor vehicle?

    a. It can generate too much power in some situations.

    b. It produces hydrogen sulfide gas.

    c. It produces high levels of nitrous oxide.

    d. It is expensive to produce, store, and transport.

9. A lead-acid battery, if improperly disposed of, can cause pollution in the form of

    a. CO gas emissions.

    b. particulate matter.

    c. heavy-metal contamination.

    d. hydrogen adsorption.

10. Which of the following units expresses the storage capacity of a NiMH battery in terms of the mathematical integral of delivered current with respect to time?

    a. Volt-hours

    b. Ampere-hours

    c. Watt-hours

    d. Kilowatt-hours

CHAPTER 9

Exotic Propulsion Methods

In this chapter, we'll look at propulsion alternatives for three modes of transportation that ought to prove interesting in the near future: trains, ships, and spacecraft. For trains, *magnetic levitation* holds promise. For ships, *nuclear power* may well make a resurgence. For spacecraft propulsion, *ion rockets*, *hydrogen-fusion engines*, and *solar sails* have been considered.

Magnetic Levitation

Magnetic levitation takes advantage of magnetic forces to suspend moving objects above fixed media. The technology is based on the fact that objects having strong magnetic poles of the same sense (that is, north-and-north or south-and-south) exhibit a mutual, powerful repulsive force when they are brought into close proximity.

EARNSHAW'S THEOREM

Imagine approximately 100 small, pellet-shaped permanent magnets glued down, evenly spaced, on the inside surface of a plastic bowl with the north poles facing upward. This forms a large magnet with a concave, north-pole surface. Suppose

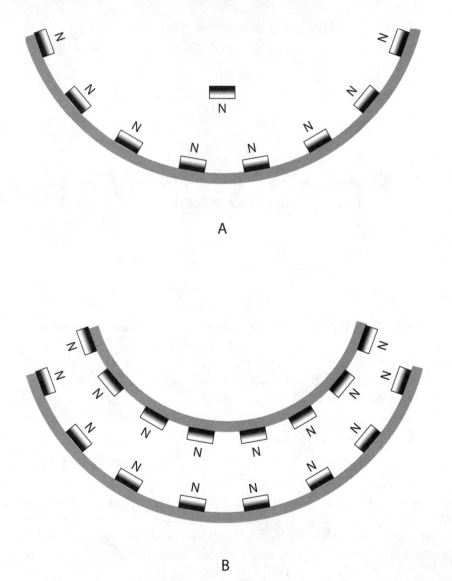

Figure 9-1   When you try to levitate a magnet above a set of other magnets as shown at A, the top magnet flips over and sticks to one of the others. Instability also occurs with two bowl-shaped magnetic structures, one above the other, as shown at B.

you anchor this bowl to a tabletop, and then take a single pellet-shaped magnet and hold it with its north pole facing downward over the center of the bowl as shown in Figure 9-1A. As soon as you let go of the single magnet, it flips over and sticks to one of the magnets inside the bowl.

Now suppose that you take another 100 magnets and glue them to the outer surface of another mixing bowl the same shape as, but somewhat smaller than, the first bowl, with the north poles facing outward, forming a large magnet with a convex north-pole surface. Suppose you try to set this bowl down inside the first one, as shown in Figure 9-1B. You hope that the top bowl will hover above the bottom one, but it doesn't. It finds some way to land, off-center, on the bottom bowl. If there are enough magnets on the bowls to prevent a landing inside the bottom bowl, the top bowl will skitter off and land outside the bottom one.

Magnetic levitation cannot be achieved with a set of static (nonmoving and nonrotating) permanent magnets. There is always instability in such a system, and this instability is magnified by the slightest disturbance. This fact was proven by Samuel Earnshaw in the 1800s, and became known as *Earnshaw's theorem*. Despite the conclusion of this theorem, it is possible to obtain magnetic levitation. Earnshaw's theorem is based on a narrow set of assumptions, and systems can be built to get around these constraints. Earnshaw's theorem applies only to systems that consist exclusively of permanent magnets with no relative motion among them. In recent years, scientists have come up with dynamic (moving) systems of magnets that can produce levitation. Some such systems have been put to use in railway trains.

FEEDBACK SYSTEMS

Consider the two-bowl scenario of Figure 9-1B. If you try this experiment, you'll never get the upper bowl to hover indefinitely above the lower bowl in free space. But suppose you build a *feedback system* that keeps the upper bowl in alignment with the lower one?

Here's a crude example of how an *electromechanical feedback system* can produce magnetic levitation with the two-bowl scheme. It has an electronic *position sensor* that produces an *error signal*, and a mechanical *position corrector* that operates, based on the error signals from the position sensor, to keep the upper bowl from drifting off center. As long as the upper bowl is exactly centered over the lower one (Figure 9-2A), the position sensor produces a zero output signal. The upper bowl has a tendency to tilt or move sideways because of instability in the system (Figure 9-2B). As soon as the bowl gets a little off center, the position sensor produces data that describes the extent and direction of the displacement. The error data has two components: a *distance error signal* that gets stronger as the off-center displacement of the upper bowl increases, and a *direction error signal* that indicates

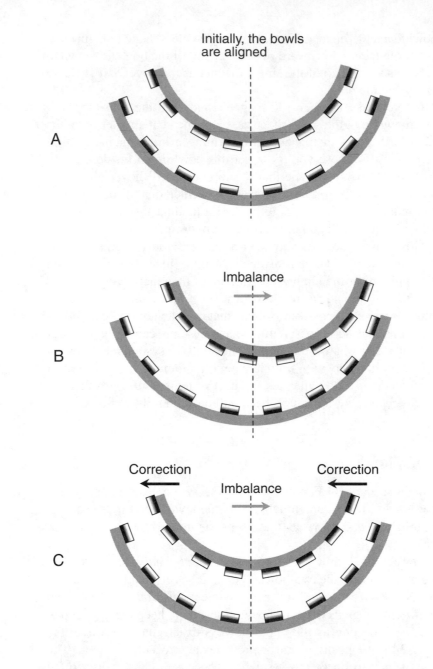

Figure 9-2   A feedback system can keep two magnet-arrayed bowls centered, producing levitation. At A, the initial aligned situation. At B, an imbalance throws the upper bowl off center. At C, a correction force re-centers the upper bowl.

the direction in which the upper bowl has drifted. These signals are sent to a microcomputer, which operates a mechanical device that produces the necessary amount of force, in exactly the right direction, to get the upper bowl back into alignment with the lower one (Figure 9-2C).

DIAMAGNETISM

Certain substances known as *diamagnetic materials* cause magnetic fields to weaken. A sample of diamagnetic material spreads out, or dilates, the *magnetic lines of flux* when brought into a magnetic field. The phenomenon of *diamagnetism* is the opposite of the more commonly observed effect called *ferromagnetism*. A *ferromagnetic material* such as iron concentrates the magnetic lines of flux when brought into a magnetic field; this makes it possible to construct electromagnets.

A diamagnetic substance such as distilled water can, under certain conditions, give rise to magnetic levitation. A diamagnetic object repels either pole of a permanent magnet, just as a ferromagnetic object (such as an iron nail) attracts either pole. This force is too weak to be noticeable with ordinary magnets, because they're not strong enough. But levitation can occur if a lightweight diamagnetic object is placed inside a bowl-shaped container arrayed with powerful electromagnets, as shown in Figure 9-3. A small drop of distilled water, for example, can be suspended in midair by this method under controlled laboratory conditions. The force thus produced is too weak to support heavy objects such as train cars. In order to get a strong repulsive force as a result of diamagnetism, another effect must be exploited: the ability of certain media to become perfect electrical conductors when they become extremely cold.

SUPERCONDUCTORS

At temperatures approaching absolute zero (about −273.15°C or −459.67°F), some metals lose all of their electrical resistance. Such an electrical medium is called a *superconductor*, and the phenomenon of zero resistivity (or perfect conductivity) is called *superconductivity*. An electrical current in a superconducting loop of wire can circulate around and around, without growing noticeably weaker, for a long time.

Superconductors make magnetic levitation possible because the magnetic flux is completely expelled from such a medium. The lines of flux are dilated so much that they disappear altogether within a superconductor. This phenomenon is called *perfect diamagnetism*. A more technical name for it is *Meissner effect*, named after one of its discoverers, Walter Meissner, who first noticed it in the 1930s.

Because of the Meissner effect, superconductors have generated interest among engineers seeking to build vehicles and other machines that take advantage of magnetic levitation. A superconducting diamagnetic medium produces a much stronger repulsive force, for a given magnetic field intensity, than any ordinary

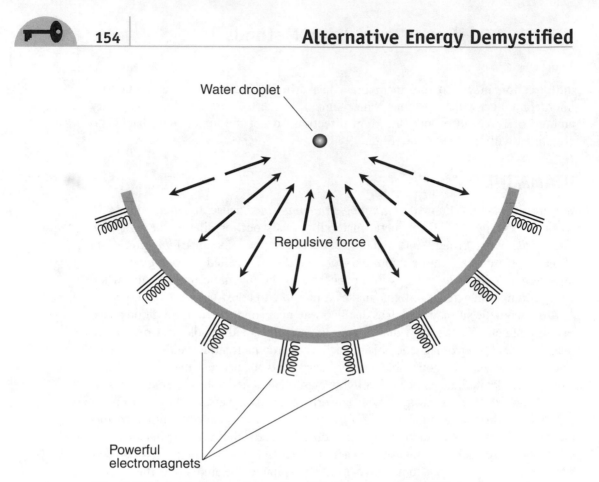

Water droplet

Repulsive force

Powerful
electromagnets

Figure 9-3 Levitation can be produced by the action of powerful electromagnets against a diamagnetic substance such as a droplet of water.

diamagnetic sample. The repulsion takes place whether the proximate magnetic pole is north or south, so instabilities associated with systems such as those in Figures 9-1 and 9-2 do not occur.

ROTATION

An arrangement similar to that shown in Figure 9-1A can be made to work if the upper magnet rotates at a rapid rate, like a hovering top. This makes it act as a stabilizing *gyroscope*. A disk made of non-ferromagnetic material (that is, a substance that does not attract magnets) is attached to the upper magnet to increase the gyroscopic effect. Figure 9-4 illustrates this scheme. When it spins fast enough, the upper magnet balances on the mutually repulsive magnetic fields produced by itself and the lower array of magnets. A toy called the *LEVITRON* provides a

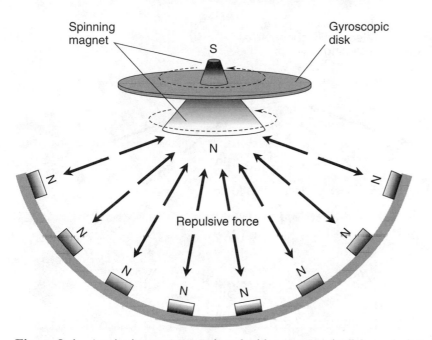

Spinning magnet

Gyroscopic disk

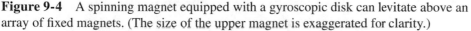

Repulsive force

Figure 9-4   A spinning magnet equipped with a gyroscopic disk can levitate above an array of fixed magnets. (The size of the upper magnet is exaggerated for clarity.)

fascinating display of this principle. An explanation of how it works can be found on the Internet at *www.levitron.com.*

A system such as the one shown in Figure 9-4 does not violate Earnshaw's theorem, which applies only to fixed magnets. Nevertheless, the system shown in Figure 9-4 is temperamental. The *angular speed* (rate of rotation) must be between certain limits, and the spinning magnet must be shaped just right. After awhile, air resistance will cause the spinning magnet to slow down to the point where the gyroscopic effect fails, and the upper magnet flips over and sticks to one of the magnets in the lower array.

OSCILLATING FIELDS

An object that conducts electric currents but does not concentrate magnetic flux, such as an aluminum disk, exhibits diamagnetism in the presence of an oscillating magnetic field. Such a field can be produced by a set of electromagnets to which high-frequency AC is applied. A suitably shaped, rotating disk made of a non-ferromagnetic metal such as aluminum will levitate above an array of such electromagnets, as shown in Figure 9-5.

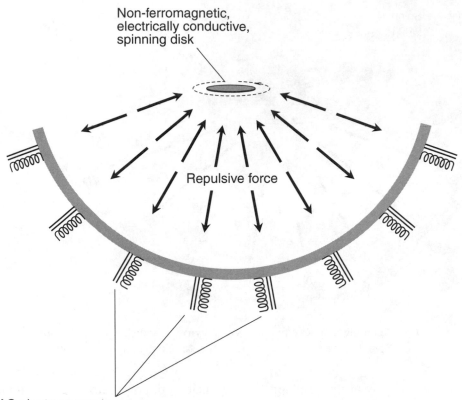

Non-ferromagnetic,
electrically conductive,
spinning disk

Repulsive force

AC electromagnets

Figure 9-5 A spinning disk, made of non-ferromagnetic material that conducts electric currents, can levitate above a set of AC electromagnets.

This method of levitation works because of *eddy currents* that appear in the conducting disk. The eddy currents produce a secondary magnetic field that opposes the primary, oscillating field set up by the array of fixed AC electromagnets. The disk becomes an "antimagnet." The rotation provides a stabilizing, gyroscopic effect, so the disk doesn't flip over. The disk stays centered as long as it is placed along the central axis of the array to begin with, and as long as there isn't a significant disturbance to throw it off center.

Eventually the rotation rate of the disk will decay, because of air resistance, to the point that the system becomes unstable. This problem can be eliminated by placing the entire system in a vacuum. (The same thing can be done with the rotating-magnet system described in the previous section.) This allows the system to operate forever in theory, although in practice it will eventually fail because of inescapable real-world friction and energy loss.

The Maglev Train

High-speed rail transit is one of the most exciting applications of magnetic levitation. Some passenger trains for urban commuters employ this technology, which is called *maglev*. The diamagnetic effects of superconductors are most often used for such systems. An adaptation of the rotation scheme, described above, has also been tested on a small scale.

HOW IT WORKS

In a maglev train, there is no contact between the cars and the track. The only friction occurs as a result of the air resistance encountered by the moving cars, which are suspended over a monorail track. There is a gap of 2 to 3 cm (about 1 in) between the train and the track.

The cars in a superconductor maglev train can be supported by either of two geometries, shown in Figure 9-6. In the scheme shown at A, the cars are attached to bearings that wrap around the track. Drawing B shows a system in which the track wraps around bearings attached to the cars. In either arrangement, vertical magnetic fields keep the cars suspended above the track, and horizontal magnetic fields stabilize the cars so they remain centered. Acceleration and braking are provided by a set of *linear motors*, which require an additional set of electromagnets in the track.

An alternative system called *Inductrack* uses permanent magnets in the cars and wire loops in the track. The motion of the cars with respect to the track produces the levitation, in a manner similar to the way a rotating, conducting disk levitates above a set of fixed magnets. This system travels on sets of small wheels as it first gets going. Once it is moving at a few kilometers per hour, the currents in the loops become sufficient to set up magnetic fields that repel the permanent magnets in the train cars. As with the superconductor type maglev train, the Inductrack system uses linear motors to achieve propulsion and braking.

ADVANTAGES OF THE MAGLEV TRAIN

- Maglev trains are capable of higher speeds than conventional trains.
- Maglev trains make less noise than conventional trains.
- Maglev trains can reduce commute times for people who use trains.
- Maglev trains can make use of low-pollution electrical energy sources.

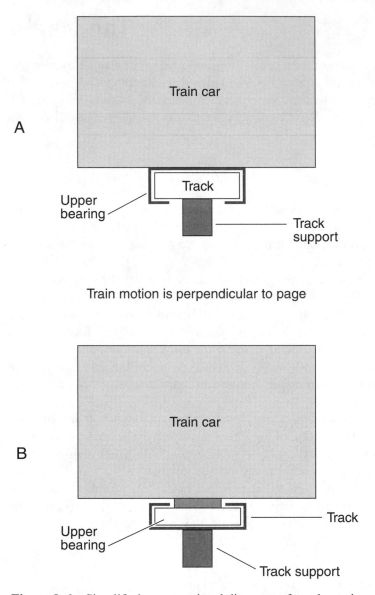

Figure 9-6 Simplified cross-sectional diagrams of maglev train geometries. At A, an upper bearing, attached to the car, wraps around and levitates above a monorail track. At B, the upper bearing levitates inside a wrap-around track. In both illustrations, train motion is perpendicular to the page.

LIMITATIONS OF THE MAGLEV TRAIN

- Maglev trains are more expensive than conventional trains.
- Special training is required for maglev maintenance personnel.
- Superconductor maglev trains rely on powerful electromagnets embedded in the track to obtain the levitation. This raises the problem of shielding passengers from the strong magnetic fields.
- A sudden power outage will cause the cars in a superconductor maglev train to settle onto the track. This may be dangerous if it occurs at high speed. (In an Inductrack train, the wheels eliminate this danger, allowing the cars to coast to a stop.)
- A high crosswind may disrupt the operation of a maglev train by de-centering the cars and causing them to contact the track. Snow or ice on the track can also cause trouble.

PROBLEM 9-1
How can passengers be shielded from the strong magnetic fields in a superconductor-type maglev train?

SOLUTION 9-1
The train cars, or at least the passenger compartments within them, can be made of a ferromagnetic substance such as steel, which tends to block magnetic lines of flux. Unfortunately, steel is much heavier than aluminum, the other metal commonly used in general construction. Aluminum is not ferromagnetic and would offer no protection against magnetic fields unless it were made to carry high and potentially dangerous electric currents.

PROBLEM 9-2
How can a maglev train negotiate a steep hill or mountain? Won't it fall downhill and settle at the bottom of a valley if there is no friction to provide braking action?

SOLUTION 9-2
The linear motors used in maglev train systems can drive the cars up steeper grades than is possible with conventional trains. In addition, the linear motors can provide braking action by switching into reverse, and they can keep the train from falling downgrade by operating against the force of gravity.

The Nuclear-Powered Ship

Nuclear-powered oceangoing vessels have existed for decades. The principal maritime application of nuclear power has historically been in submarines and aircraft carriers. However, with the future of the world's oil supply in doubt, *uranium fission* has become a subject of interest for propelling large marine vessels, both military and commercial.

HOW A FISSION-POWERED SHIP WORKS

In the type of *nuclear reactor* in use today, elemental uranium gradually decays into lighter elements. The term *nuclear* derives from the fact that the atomic *nucleus* is involved. In fission, the uranium nuclei, consisting of *protons* and *neutrons*, are split apart. Thus, a heavy element is transformed into lighter elements—an ancient alchemist's dream come true!

As fission takes place, energy is released in the form of heat and *ionizing radiation*, also known as *radioactivity*. Some of this radiation consists of high-speed neutrons that break more of the uranium atoms apart, releasing still more energy. If the reaction occurs rapidly, an explosion occurs. This is the principle by which the first atom bombs worked. However, the chain reaction can be slowed down and made self-sustaining. When this is done under rigidly controlled conditions, uranium fission can provide large quantities of usable heat energy for long periods of time, and the risk of explosion is essentially nil.

Figure 9-7 is a block diagram of the power plant in a typical nuclear-powered ship or submarine. Heat from the reactor is transferred to a *water boiler* by means of *heat-transfer fluid*, which bears some resemblance to the fluid used in heat pumps, air conditioners, and automobile radiators. This heat-transfer fluid, also called *coolant*, passes from the shell of the boiler back to the reactor through a *coolant pump*. The water in the boiler is converted to steam, which drives a turbine. After passing through the turbine, the steam is condensed and sent back to the boiler by a *feed pump*. The water and heat-transfer fluid are entirely separate, closed systems; neither comes into direct physical contact with the other. This prevents the accidental discharge of radiation into the environment through the water/steam system.

The turbine can turn the propeller through a *drive system*. The turbine also turns the shaft of a generator that provides electricity for the crew, passengers, and electronic systems. The electricity from this generator can also power a motor, or charge a battery that powers a motor, connected to the drive system. This electric motor can provide supplemental or backup propulsion.

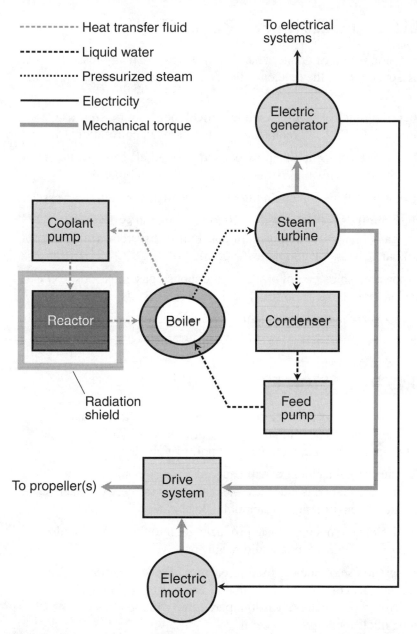

Figure 9-7   Simplified block diagram of a nuclear power and propulsion system suitable for use in large oceangoing vessels.

ADVANTAGES OF FISSION FOR PROPULSION

- Nuclear fission provides greater "nautical mileage" than any fossil fuel per unit mass, even taking the reactor radiation shield (which is massive) into account.

- Maintenance of nuclear reactors, while critical, need not be done as frequently as refueling and maintenance operations in conventional vessels.

- Nuclear-fission reactors and their associated peripherals can operate in the absence of oxygen. This makes them ideal for use in submarines.

- Nuclear-powered ships can attain higher speeds than conventional vessels.

- When the overall dangers of nuclear-fission propulsion systems are weighed against the proven dangers of conventional systems (oil spills, for example), nuclear fission compares favorably.

- Nuclear-powered ships do not produce greenhouse-gas emissions, CO gas, or particulate pollutants as do fossil-fuel-powered vessels.

- Nuclear energy in general can help the world economy wean itself off of fossil fuels.

LIMITATIONS OF FISSION FOR PROPULSION

- Nuclear fission reactors produce certain waste products that remain radioactive for many years (although most of it decays in a few months). Disposal processes present technical and political challenges.

- Although the risk of accident or sabotage involving a nuclear reactor is small, the potential consequences—leakage of extremely radioactive material into the environment—cannot be discounted.

- If certain fissionable nuclear waste products get into the wrong hands, nuclear terrorism or blackmail could result.

- The widespread use of nuclear fission reactors faces opposition from certain groups because of the above-mentioned negative factors. This has given rise to public apprehension, particularly in the United States, concerning nuclear energy in general.

PROBLEM 9-3

Why can't nuclear waste be dumped straightaway into the sea? The world's oceans are vast, and the quantity of nuclear waste, even if there are lots of reactors in use, will be small. Wouldn't the oceans dilute the radiation to negligible levels? If that isn't acceptable, why not put the waste in sealed containers with radiation shielding and drop them to the bottom of the sea, far from land masses or populated islands?

SOLUTION 9-3
Nuclear waste has a high radiation output in proportion to its mass. Besides this, it is difficult to say how low levels of radiation, distributed throughout the world's oceans, might affect marine life and disrupt the food chain from plankton all the way up to humanity. Sealed containers would have to be exceptionally well-made so they would not rupture before the radioactivity had decayed to safe levels.

The Ion Rocket

Hot gases produced by the combustion of flammable fuels, as in conventional rockets, are not the only way to produce thrust for spacecraft. Another way to propel interplanetary and interstellar spacecraft makes use of powerful *linear particle accelerators*. Atomic nuclei are accelerated to high speed and ejected out the rear of the spaceship, resulting in forward impulse according to the principle of action/ reaction.

HOW IT WORKS

Figure 9-8 is a simplified functional diagram showing how an *ion rocket engine* can work. The source ejects large quantities of ionized gas such as hydrogen or helium. Positive ions of simple hydrogen consist of individual protons. Positive ions of helium usually consist of two protons and two neutrons. Any atomic nucleus is, in fact, a positive ion that can be accelerated by negatively charged *anodes*, through which the nuclei pass. The voltages on the anodes are extremely high, creating powerful electric fields. Successive anodes are supplied with higher and higher negative voltages.

As the ions travel through anode after anode, the particles gain speed, and consequently they gain momentum. When the ions pass through the last anode, they are moving so fast, and have so much rearward momentum (the product of mass and speed), that the reaction force pushes the spacecraft forward. This force is not great, but it is sustained. Over a long period of time in the near-vacuum of deep space, a spacecraft of this sort can theoretically reach speeds far greater than those attainable by conventional rocket-powered vessels.

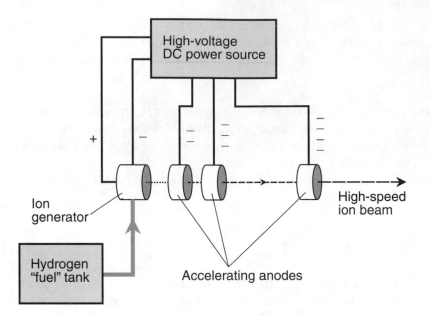

Figure 9-8   Conceptual diagram of an ion rocket engine that produces thrust by accelerating atomic nuclei (ions) to high speed.

ADVANTAGES OF THE ION ROCKET

- Ion engines are efficient. They utilize most of the input energy to produce thrust.

- An ion engine can keep operating for a long time, and thereby can allow a small vessel to achieve high speed, although the rate of acceleration is slow.

- Ion rockets are inherently safe, because the fuel does not have to be carried in a form that could combust or explode in outer space.

LIMITATIONS OF THE ION ROCKET

- Linear particle accelerators require large amounts of power in order to function. The only currently existing technology that can provide the necessary power over the required period of time is nuclear fission, in the form of an onboard reactor.

- For long journeys, obtaining and carrying fuel (hydrogen or helium) in the required large amounts could be a problem.

- Because an ion rocket accelerates slowly, it cannot be used as the launch vehicle to put a spacecraft into earth orbit. It is workable only for vessels that are already in space.

PROBLEM 9-4

What factor determines the highest speed attainable by an ion engine operating indefinitely in interstellar space? How is this different from the factor that determines the highest speed attainable by a jet aircraft operating in the earth's atmosphere?

SOLUTION 9-4

In a jet-powered aircraft, the maximum attainable speed is equal to the speed of the ejected matter relative to the engines. Air resistance prevents the craft from moving forward any faster than the exhaust is hurled out rearward. But in space, the situation is different. As long as thrust is maintained, speed can increase without limit. The thrust from an ion engine depends on the momentum of the ejected matter. If protons are ejected at a constant speed and a constant rate from the rear of a spaceship, the thrust remains constant, and the acceleration therefore also remains constant. Given enough time, even modest acceleration, if maintained, can result in extreme speed. The theoretical upper limit is the speed of light, c, according to the *theory of special relativity*. That's approximately 300,000 km/s or 186,000 mi/s.

Fusion Spacecraft Engines

Hydrogen fusion is the nuclear process in which atoms of hydrogen combine, at extremely high temperatures, to form atoms of helium. In this process, energy is liberated. Hydrogen fusion is far more efficient than uranium fission in converting matter to energy. This is the type of reaction believed to take place deep inside the sun and most other stars.

HOW THEY WORK

Several types of fusion-powered spacecraft have been proposed by scientists and aerospace engineers as alternatives for obtaining the speeds necessary for long-distance space journeys, using fuel that would be of reasonable mass. These include designs known as the *Orion*, the *Daedalus*, and the *Bussard ramjet*.

   In the Orion spaceship, hydrogen fusion bombs would be exploded at regular intervals to drive the vessel forward. The force of each blast, properly deflected, would accelerate the ship. The blast deflector would be strong enough to withstand the violence of the bomb explosions, and it would be made of material that would not melt, vaporize, deform, or erode because of the explosions. The blast deflector would also serve as a radiation shield to protect the astronauts in the living quarters. This idea has been all but rejected in recent years because of technical problems, such as the extreme bursts of acceleration (harmful to the crew) and the danger that the bomb blasts could damage or disable the ship.

A smoother ride would be provided by the Daedalus design. This would replace the bombs with a *fusion reactor* that would produce a continuous, controlled "burn." This vessel would use a blast deflector similar to that used in the Orion design. The advantage of the Daedalus ship would be that the acceleration would be steady, rather than intermittent. Daedalus could attain high speed without subjecting the astronauts to bursts of extreme acceleration.

Both the Orion and the Daedalus ships could, according to their proponents, reach approximately 10 percent of the speed of light ($0.1c$). The ship could then travel to and from the sun's nearest stellar neighbors in a few human lifespans. The ships would accelerate for the first half of the journey, and decelerate for the second half. Deceleration would be obtained by rotating the vessel 180° so the exhaust would be directed forward, producing a rearward impulse.

The most intriguing nuclear fusion design is the Bussard ramjet. This is similar to the Daedalus, but it would not have to carry nearly as much fuel. Once the ship got up to a certain speed, a huge "scoop" in the front could (hopefully!) gather up enough hydrogen atoms from interstellar space to provide the necessary fuel for hydrogen fusion reactions. Figure 9-9 is a simplified diagram of this type of spaceship. The greater the speed attained, the more hydrogen the ship could sweep up, thus helping it go even faster. The ship would decelerate by directing the exhaust out of a forward-facing nozzle in the center of the scoop.

The Bussard ramjet would work especially well in gaseous nebulae, provided those regions were not too peppered with meteoroids and asteroids. Some scientists think that this type of spaceship could attain speeds great enough to take advantage of a phenomenon called *relativistic time dilation*. The astronauts on board the ship would age more slowly than they would if the vessel were traveling at conventional speeds, and this effect could increase the distance reachable within a human lifetime. Time dilation becomes significant at speeds greater than 90 percent of the speed of light ($0.9c$)—roughly 270,000 km/s or 170,000 mi/s. As the speed approaches c, the time-dilation factor increases without limit.

ADVANTAGES OF FUSION SPACECRAFT ENGINES

- Hydrogen is the most abundant element in the universe. It exists in free form in outer space. This could provide an unlimited supply of fuel, or at least minimize the amount that would have to be carried.

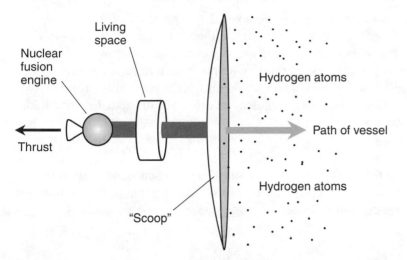

Figure 9-9   The Bussard ramjet gets its fuel by scooping up hydrogen atoms from space and using them to fuel a fusion engine.

- Hydrogen fusion produces a tremendous amount of energy from a small amount of matter.

- Hydrogen fusion produces essentially no radioactive waste. The only significant byproduct is helium. No CO, CO_2, SO_x, NO_x, or particulate pollutants are generated either.

- A fusion reactor could provide all the energy necessary to operate the electrical, electromechanical, and electronic systems on board the vessel.

LIMITATIONS OF FUSION SPACECRAFT ENGINES

- The crew would have to be protected from the radiation produced by the fusion reaction. This would necessitate massive shielding, reducing the attainable acceleration for a given amount of thrust.

- Even though a fusion engine may allow a vessel to attain high speed, it is unrealistic to suppose that humans will ever use such spaceships to "roam the galaxy" as they do in science-fiction movies and television shows. Some entirely different means of propulsion, as yet unknown, will have to be developed in order for that to be possible.

- The materials in the blast deflector would have to be capable of withstanding extreme temperatures, as well as immense mechanical stresses, for a prolonged period of time.

- If near-light speed were attained to take advantage of time dilation, the earth would age more rapidly than would the vessel's crew. If the astronauts were to return to earth after a long journey of this sort, they would find themselves in the distant future with no "return ticket." The prior knowledge that this phenomenon would take place could be disturbing to would-be travelers, and the psychological effects of the actual event could be devastating.

PROBLEM 9-5

Could a hydrogen-fusion-powered space vessel be launched directly from the earth's surface? If so, wouldn't the blast from the engines cause destruction and deadly radiation near the launch site?

SOLUTION 9-5

Most proposals for hydrogen-fusion-powered spacecraft envision a conventional rocket launch vehicle to place the ship in an earth orbit. The fusion engines would be activated at an altitude of several thousand kilometers. Thus, there would be no dangerous effects on the earth's surface.

The Solar Sail

The sun emits some of its energy in the form of high-speed subatomic particles. These particles create a so-called *solar wind* that rushes radially outward past the planets. It ought to be possible to take advantage of this solar wind to propel a spaceship, in much the same way as water-sailing vessels are propelled by moving air molecules on earth.

HOW IT WORKS

A spaceship with a *solar sail* would require an enormous sheet of reflective fabric or foil. This sheet would be attached to the living quarters (see Figure 9-10). The range of travel would be limited to the solar system (or whatever other star system the travelers might happen to be visiting). It would be easier to travel outward away from the star than inward toward it, for obvious reasons. But, just as water-sailing vessels can make progress into the wind by *tacking* (taking a zigzag path), "solar sailors" would be able to navigate in any direction, given sufficient solar wind speed. To move in closer to the sun, they would follow a spiraling path inward, the direction of travel subtending an angle of slightly less than 90° with respect to the solar wind.

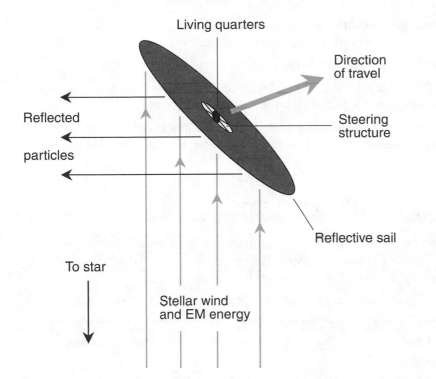

Living quarters

Direction
of travel

Reflected

particles

Steering
structure

Reflective sail

To star

Stellar wind
and EM energy

Figure 9-10   A spaceship with a solar sail rides on solar wind, just as a sailboat rides on
atmospheric wind.

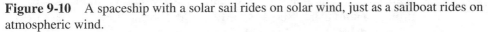

The solar sail requires no onboard fuel, at least in the ideal case. However, navigation
would be tricky. Sudden *solar flares* would produce a dramatic increase in the solar
wind because of the large number of relatively massive particles ejected during such an
event. Near any planet with a magnetic field, including the earth, the particles are
deflected and the solar wind does not necessarily "blow" away from the star.

The classic ocean sailing ships did not come in all the way to the beach, but
dropped anchor in deep water and sent small boats to shore. During space voyages,
the main ship could draw in its solar sail and "drop anchor" by falling into a planetary
orbit. Landings could be made with small shuttles that would be sent out by, and
return to, the main ship.

ADVANTAGES OF THE SOLAR SAIL

- The solar sail is a passive device. It does not require the ship to carry, or
  collect, any fuel. (However, a backup propulsion system, such as a set of
  conventional rockets or a nuclear-fusion engine, would be a good idea!)

- The mechanical and thermal stresses on a solar sail are far smaller than those in a conventional rocket engine or a nuclear-fusion engine.

- A solar sail produces absolutely no radiation or waste products.

- Once deployed, a solar sail would require little or no maintenance, except for the occasional repair of punctures produced by meteoroids.

- A solar sail could be partially coated with thin-film photovoltaics (solar panels) to provide electrical energy for onboard hardware and life-support systems.

LIMITATIONS OF THE SOLAR SAIL

- The solar sail relies on the stream of particles from the sun (or a nearby star). It might not work in interstellar space for lack of a dominant star to produce a defined and predictable particle wind.

- The maximum speed attainable with a solar sail would be much lower than the maximum speed of a spaceship powered by an ion or fusion engine.

- A solar sail, because of its large size, would be difficult and awkward to deploy.

- Sudden solar flares and planetary magnetic fields would complicate the navigation of a ship using a solar sail. It would be something like sailing a marine vessel through a region with shifting winds and ocean currents.

PROBLEM 9-6
Wouldn't the huge surface area of a solar sail create friction with the rarefied gases normally present in interplanetary space, rendering the system impracticable?

SOLUTION 9-6
There would be some friction between the solar sail and atoms of matter—mainly hydrogen and other gases—in space. However, the force produced by the solar wind should be far greater than the resistance offered by this friction, at least in the inner solar system. But of course, we'll never know how well a solar sail will work until it is tested "in the field"!

Quiz

This is an "open book" quiz. You may refer to the text in this chapter. A good score is eight correct. Answers are in the back of the book.

1. In an oceangoing ship powered by a nuclear reactor, the electric generator derives its torque (rotational force) directly from
    a. the reactor core.
    b. the heat-transfer fluid.
    c. the water boiler.
    d. a steam-driven turbine.

2. The energy for a solar sail would come from
    a. electrostatic charges.
    b. moving particles.
    c. nuclear fusion.
    d. nuclear fission.

3. The CO gas emission from a nuclear-fusion spacecraft engine
    a. could cause harmful ionization of the earth's atmosphere immediately after launch.
    b. would be considerable, but not of concern because it would disperse in space.
    c. could, if not diverted from the spaceship, be dangerous to the crew.
    d. is zero; there is no CO gas emission from a nuclear-fusion spacecraft engine.

4. The momentum of a proton ejected from an ion engine is equal to
    a. the speed of the proton times its mass.
    b. the speed of the proton divided by its mass.
    c. the mass of the proton times the speed of light.
    d. the mass of the proton divided by the speed of light.

5. Magnetic lines of flux are concentrated as a result of a phenomenon called
    a. diamagnetism.
    b. superconductivity.
    c. Meissner effect.
    d. ferromagnetism.

6. A sample of diamagnetic material is attracted to

   a. a north magnetic pole.

   b. a south magnetic pole.

   c. either a south or a north magnetic pole.

   d. neither a south nor a north magnetic pole.

7. In a fission reactor,

   a. heavy atomic nuclei are split, forming lighter atomic nuclei.

   b. light atomic nuclei merge, forming heavier atomic nuclei.

   c. ions are accelerated to high speed to produce propulsion.

   d. the powerful magnetic fields can be dangerous at close range.

8. In a maglev train, the only significant friction occurs between

   a. the cars and the track.

   b. the cars and the electromagnets.

   c. the cars and the atmosphere.

   d. the opposing magnetic fields.

9. A significant advantage of an ion engine for spacecraft propulsion is the fact that it

   a. can produce a great deal of thrust.

   b. converts most of the available energy into thrust.

   c. does not require an onboard source of power.

   d. produces only helium as a byproduct.

10. Earnshaw's theorem, which seems at first glance to rule out any possibility of achieving magnetic levitation in the real world, applies specifically to

   a. magnetic fields of opposing polarity.

   b. sets of fixed, permanent magnets.

   c. sets of electromagnets in motion.

   d. sets of rotating magnets.

CHAPTER 10

Electricity from Fossil Fuels

As nonrenewable sources of electrical energy become less abundant and more costly, the nations of the world will have no choice but to exploit alternatives. But for now, a significant proportion of the world's electricity comes from generators driven by turbines that ultimately get their energy from fossil-fuel combustion.

Coal-Fired Power Plants

Although coal is a nonrenewable resource, there is still plenty of it in the ground. In the first years of the 21st century, coal experienced a resurgence in usage as supplies of oil and natural gas, the other combustibles most often used to generate electricity, grew short. Coal is not a permanent solution to humankind's energy problems, but it's there for our responsible use as we work toward developing more-lasting alternatives.

HOW THEY WORK

The very idea of a coal-fired power plant sounds primitive to the uninformed person. However, a modern coal-burning electric-generating facility is a sophisticated operation. Coal is mined in locations that are usually some distance away from generating plants, so the coal must be transported. This is most often done by rail. The coal is cleaned and *de-gassed* at the mining site prior to being loaded onto *coal trains* that trundle endlessly across the countryside.

Figure 10-1 is a simplified functional diagram of a coal-fired electric-generating plant. Once the coal arrives at the generating site, it is loaded into a large *hopper*

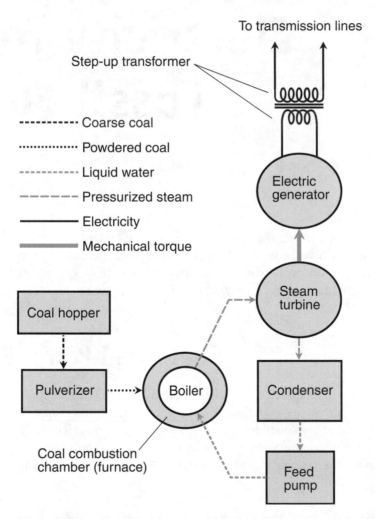

Figure 10-1   Simplified functional block diagram of a coal-fired electric-generating plant using a boiler and steam turbine.

(although excess may be simply piled up, forming little black hills at the site). From the hopper, the coal moves into a *pulverizer* that grinds it into a powder as fine as dust. A *blower* (not shown) forces air laden with coal dust into the *combustion chamber*, where the coal dust is burned. The resulting heat turns liquid water into pressurized steam in the *boiler*. The water supply is purified to get rid of minerals that would otherwise build up in the system over time. The steam drives a *steam turbine*, which turns the shaft of one or more *generators*. The steam from the turbine is cooled and condensed back into liquid water by a *condenser*. A *feed pump* returns this liquid water to the boiler.

The electricity produced by the generator is AC at a frequency of 60 hertz (Hz) in the United States. In some countries the frequency is 50 Hz. One hertz is equivalent to one complete wave cycle per second. The voltage of this AC is increased to several hundred thousand volts by a huge *step-up transformer* connected to each generator. The high-voltage electricity is then fed to the *transmission lines* for distribution.

WHY SUCH HIGH VOLTAGE?

Have you ever wondered why long-distance power transmission is done at high voltage, necessitating massive towers and gigantic insulators? Why can't electricity be transmitted at low voltage in heavy-duty wires running between modest structures, or even underground? There's a good reason.

For a given amount of power ultimately consumed by the *load* (the set of all the end users), the current in a transmission line goes down as the voltage goes up. Reducing the current reduces the *power loss* in an electrical transmission line. In order to understand why, recall from your basic electricity course the following formula:

$$P = EI$$

where P is the power in watts, E is the voltage in volts, and I is the current in amperes. By rearranging this formula, we can see that at any given power level, the current is inversely proportional to the voltage:

$$I = P/E$$

The power loss in a transmission line is proportional to the *square* of the current. This loss represents power that never reaches the end users; it simply heats up the transmission line. The following relationship holds:

$$P = I^2 R$$

where P is the power in watts, I is the current in amperes, and R is the resistance of the wire in ohms. Engineers can't do much about the wire resistance or the power consumed by the load, but they can maximize the voltage, and thereby minimize the current that a transmission line is forced to carry for a certain power demand.

Suppose the voltage delivered into a transmission line is stepped up by a factor of 10, and the combined loads at the end of the line draw constant power. The increase in voltage reduces the current to 1/10 of its previous value. As a result, the power loss is cut to $(1/10)^2$, or 1/100, of its former level! Obviously, the use of a step-up transformer in a single location is cheaper and easier than stringing up 100 times as much wire mass in a transmission line spanning many kilometers.

Is it frightening to contemplate a power transmission line that carries, say, 500,000 V AC? Perhaps. But the health risk from power lines (the actual extent of which is subject to debate) comes from the *magnetic fields* they generate. The strength of these fluctuating magnetic fields is proportional to the current, not the voltage. If that big transmission line through the outskirts of your town operated at 500 V rather than 500,000 V, the magnetic field near it would be far more intense, and the potential for harmful effects correspondingly greater.

ALONG THE LINE

Extreme voltage is good for *high-tension* (meaning high-voltage) power transmission, but it's of no direct use to an average consumer. The wiring in a high-tension system must be done using precautions to prevent *arcing* (sparking) and short circuits. Personnel must be kept at least several meters away from the wires. Can you imagine trying to use an appliance, say a home computer, directly with a 500,000-V electrical system? You'd be killed before you could get the plug into the socket.

At various points near groups of end users, medium-voltage power lines branch out from the major lines, and *step-down transformers* are used at the branch points. The step-down transformers reduce the voltage in the power lines. The lower-voltage lines fan out to more step-down transformers, leading to even lower-voltage lines. Finally, several lines from each transformer serve individual buildings. Each transformer must be made with wires heavy enough to easily withstand the maximum electrical current demanded by all the consumers it serves.

Sometimes, such as during a heat wave, the demand for electricity rises above the normal peak level. This loads down the circuit to the point that the voltage drops several percent. This is called a *brownout*. If consumption rises further still, a dangerous current load is placed on one or more intermediate power transformers. Circuit breakers in the transformers protect them from destruction by opening the circuit. Then there is a temporary *blackout*.

At individual homes and buildings, transformers step the voltage down to either 234 V or 117 V. Usually, 234-V electricity is provided in the form of three sine waves, called *phases*, each appearing at one of the three slots in the outlet (see Figure 10-2A). This voltage is commonly employed with heavy appliances, such as

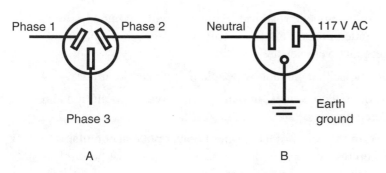

Figure 10-2   At A, an outlet for three-phase AC. At B, a conventional single-phase AC utility outlet.

electric ovens and stoves, electric furnaces, and electric laundry washers and dryers. A 117-V outlet supplies single-phase AC voltage between two of the three slots in the outlet. The third opening leads to an *earth ground* (see Figure 10-2B).

ADVANTAGES OF COAL FOR ELECTRIFICATION

- Coal is abundant, and plenty is available in the United States. This is of obvious benefit to the American economy in the face of uncertain supplies and unpredictable prices for oil and methane.

- Modern coal-fired power plants are efficient, and produce much less pollution than their "olden-days" counterparts did.

- The furnaces in power plants that use pulverized coal are flexible. They can burn all grades of coal, from lignite (soft coal) to anthracite (hard coal), and they also allow for the combustion of oil and/or methane.

LIMITATIONS OF COAL FOR ELECTRIFICATION

- The supply of usable coal, while vast, is not infinite. At best, it can provide temporary relief from, but not a permanent solution to, the world's long-term electrification problems.

- The combustion of coal, while cleaner than it used to be, generates CO_2 (a known greenhouse gas), CO, sulfur oxides (SO_x), nitrogen oxides (NO_x), and mercury compounds. Emission-control devices mitigate the air pollution when properly employed, but in countries with emerging economies, these devices are not always used.

- Coal mining leaves long-lasting scars on the landscape, and can result in runoff of toxic substances such as lead, mercury, and arsenic.

- Frequent coal trains impede road traffic in cities without enough railway overpasses or underpasses. This is not only a nuisance, but a potential danger if emergency vehicles are affected.
- Trains produce noise that can annoy people living near the tracks.
- Rail transport of coal consumes fuel, in effect increasing pollution and reducing efficiency.
- The water used in the boiler of a coal-fired power plant accumulates pollutants. When this water is replaced, the pollutants must be safely disposed of, increasing the cost of operation.

PROBLEM 10-1

Why can't huge coal-fired power plants be constructed where the mines are located, doing away with the need for coal trains?

SOLUTION 10-1

In theory, this could be done. However, much of the available coal is located far from major population centers. This would require extremely long transmission lines. The cost of constructing, operating, and maintaining these transmission lines would exceed the cost of running coal trains to smaller power plants located closer to the end users.

Oil-Fired Power Plants

In the United States, oil is used mainly for heating and propulsion, and is not widely used to generate electricity. In some respects, oil can be thought of as liquid coal. It is pumped out of the ground rather than dug or scraped out, but it must be transported from the places where it is obtained to the places where it is refined, and thence to the places where it is burned.

HOW THEY WORK

Three configurations are used in oil-fired power plants: the *conventional steam system*, the *combustion-turbine system*, and the *combined-cycle system*. In all instances, oil is transported to the generating plant from refineries, and is usually stored in tanks onsite. Transportation from the refineries to the power plant can be done in large waterborne oil tankers, by means of trains or trucks, or by means of pipelines.

In a conventional steam oil-fired power plant, the fuel is burned in much the same way as is done in an oil-fired home heating furnace, but on a larger scale. The heat from the combustion boils water. The resulting steam drives a turbine (see Figure 10-3). Except for the physical nature of the fuel, this system is similar to a coal-fired power plant.

In a combustion-turbine oil-fired power plant, the burning of the fuel produces fast-moving exhaust that passes through a *gas turbine*, which resembles a windmill designed for extreme airspeeds. The turbine drives the shaft of an electric generator (see Figure 10-4).

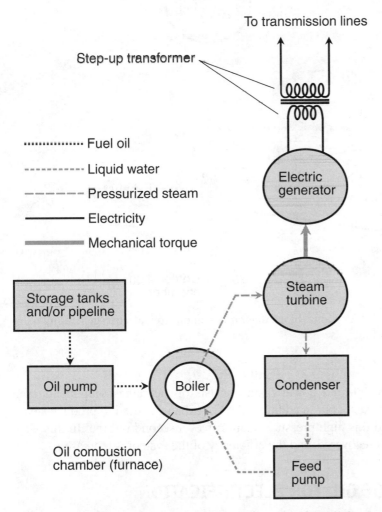

Figure 10-3　Simplified functional block diagram of an oil-fired electric-generating plant using a boiler and steam turbine.

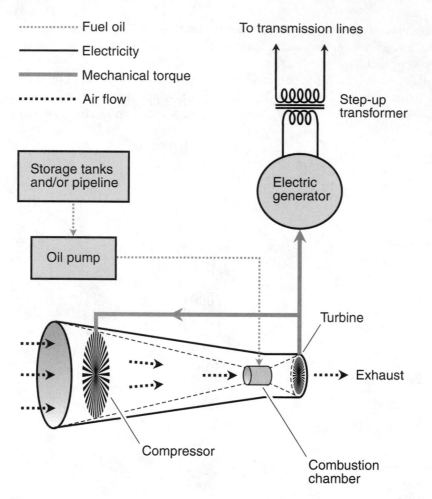

Figure 10-4 Simplified functional block diagram of an oil-fired electric-generating plant using a combustion turbine.

A combined-cycle oil-fired power plant consists of the same components as a combustion-turbine system, but in addition, the hot exhaust is used to produce steam in a water boiler, and this high-pressure steam drives a second turbine. In this way, more of the energy is recovered, and the efficiency of the system is increased.

ADVANTAGES OF OIL FOR ELECTRIFICATION

- Oil is a relatively safe fuel. An oil leak or spill can cause a fire, but it does not pose the danger of an explosion as do flammable gases.

- Oil is a high-density fuel. A moderate-sized onsite oil tank can hold enough fuel to produce a large amount of electrical energy.
- Fuel oil can be mixed with biofuel. Most oil-fired power plants can be designed to efficiently burn such a mixture, known as *hybrid fuel.*
- Steam-turbine oil-fired power plants can be designed or modified so they can also burn coal and/or methane if necessary.
- Modern oil-fired power plants produce less pollution than they did in the "olden days." This is largely because of the introduction and use of emission-control hardware in the exhaust systems.

LIMITATIONS OF OIL FOR ELECTRIFICATION

- Fuel-oil combustion, although cleaner than it once was, produces air and water pollutants similar to those produced by the combustion of coal. Emission-control systems can help reduce this pollution, but only if they are kept in proper working order. In some emerging countries, emission control is not affordable, and the result is increased pollution and greenhouse-gas production.
- The price of fuel oil is directly related to the price of crude oil. This price can spike rapidly and can be expected to rise over the long term.
- Much of the world's crude oil comes from politically unstable parts of the world, so there is an ongoing risk of sudden and unpredictable supply disruptions.
- Temporary reductions in the oil supply can result from natural events such as hurricanes, earthquakes, and pipeline corrosion.
- Oil leaks and spills can harm the environment.
- The transport of crude and refined oil by ship, rail, and truck consumes energy. This, in effect, reduces the efficiency of the whole process.
- The world's supply of crude oil is finite and nonrenewable.

PROBLEM 10-2
Do combustion type power plants require cooling systems to keep the components from overheating? Does the heat discharged from such a power plant have environmental effects?

SOLUTION 10-2
The answer to both questions is "Yes." In the interest of simplicity, cooling systems are not shown in the block diagrams here. Combustion type power plants are located

near bodies of water in order to provide a plentiful source of "coolant." (Ocean water must be desalinated for use in cooling systems.) The warm water discharged from the cooling system is eventually returned to its source, raising the temperature of the lake, river, or ocean. This affects aquatic and marine life in the vicinity of the power plant, but this is not always bad. For example, if a power plant next to a northern river keeps the river open in winter, wildlife may flock to the area during the cold season.

Methane-Fired Power Plants

Methane is a component of natural gas, a nonrenewable fossil fuel. Underground reservoirs of natural gas are plentiful but finite. In the late 20th century, methane began to replace coal, oil, and nuclear fission on a large scale for the purpose of electric power generation in the United States. Reliance on methane is expected to increase during the first part of the 21st century. But supply and price problems have been experienced with methane, just as has been the case with oil.

RECOVERY AND TRANSPORTATION

Natural gas is recovered from underground reserves and refined to obtain methane. Another way to get methane is to extract it from coal beds. Because North America has vast coal reserves, this technology can help increase the supply of methane in the medium term. In the United States, proposals to exploit coal beds for methane production encounter opposition from environmental groups and some local residents. Other residents welcome such development, because of the potential benefits to their local and regional economy.

There's still another way to get methane, and that is from the decomposition of certain biological substances such as animal waste. Methane from *biomass* contributes relatively little to the overall supply at the time of this writing, but it is a renewable resource.

The methane, once extracted and refined, can be transported to generating plants by pipelines. Alternatively, the gas can be liquefied for transportation by rail and truck, and for onsite storage in tanks. The transport of methane in any form is more dangerous than the transport of coal or oil. Coal transport can be a nuisance and an oil spill can contaminate soil and drinking water, but a methane leak can cause a deadly explosion and flash fire.

THE COMBINED-CYCLE SYSTEM

Coal-fired steam-boiler power plants can be modified to burn methane as well as oil and coal. But combustion-turbine and combined-cycle electric-generating systems are more commonly used with methane. A well-designed combined-cycle methane system can be far more efficient than older systems using boilers and steam turbines.

Figure 10-5 is a simplified block diagram of a combined-cycle electric-generating plant that burns methane. The fuel is supplied to a combustion chamber in a machine

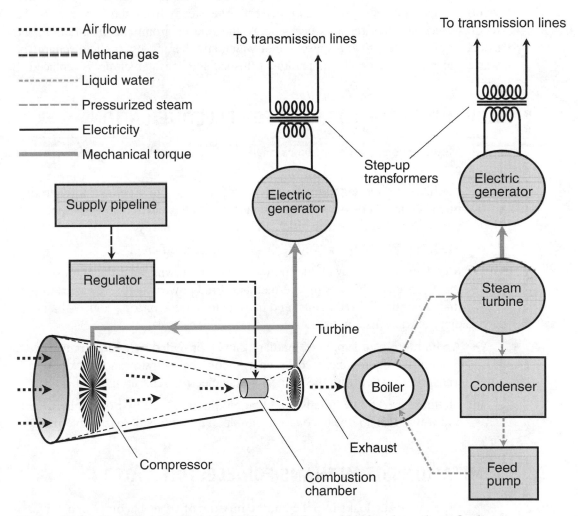

Figure 10-5   Simplified functional block diagram of a combined-cycle methane-fired electric-generating plant.

that resembles a gigantic jet aircraft engine. A *compressor* drives air into the system. The methane combustion heats this air, causing high-speed exhaust to pass through a gas turbine. The gas turbine supplies power to keep the compressor running, and also drives the shaft of an electric generator. The exhaust supplies heat energy to convert water into steam in the boiler. The steam passes through a steam turbine that drives a second electric generator.

By taking advantage of the energy contained in the exhaust from the combustion turbine, rather than simply letting it escape into the atmosphere, efficiency is optimized and *heat pollution* is reduced. The process resembles (but is a more sophisticated version of) the energy re-use that takes place in the exhaust converters of modern wood stoves. But the system isn't perfect. The steam from the boiler must be recondensed, and this releases heat energy into the environment. In addition, the steam-turbine system requires a source of water, and there is a periodic discharge of waste water when the cooling apparatus is flushed and its water supply is replaced.

ADVANTAGES OF METHANE FOR ELECTRIFICATION

- Methane combustion produces relatively low particulate, NO_x, SO_x, and CO emissions.

- Methane can serve as a medium-term transitional fuel as electric-generating plants evolve to take increasing advantage of alternative resources such as wind, solar, hydroelectric, tidal, and geothermal energy.

- Methane-fired combined-cycle power plants are efficient.

- Methane is readily available in most cities and towns, near the end users of the electricity. This means that methane-fired power plants are easier to site than coal- or oil-fired plants, and there is less need for long high-voltage electric transmission lines.

- An uninterrupted supply of methane can be provided by underground pipelines, reducing the need for energy-consuming trains and trucks to transport the fuel from the refinery to the electric-generating plant.

- Methane-fired power plants might be modified to burn hydrogen gas, when and if that fuel source becomes available in quantity at a reasonable price.

LIMITATIONS OF METHANE FOR ELECTRIFICATION

- When methane leaks into the air, the mixture of gases becomes explosive. In most locales, methane gas is given an artificial scent that is easy to recognize. This can alert people to the existence of gas leaks.

- In recent years, price "spikes" and supply disruptions have marred the reputation of methane as a reliable source of energy.

- The world's supply of naturally occurring methane is finite, and it is not renewable. (However, methane derived from biomass is renewable.)

- Exploration for, and recovery of, natural gas can adversely impact the environment by causing erosion, accelerating runoff, and increasing the risk of mudslides and floods.

- If not responsibly done, exploration for, and recovery of, natural gas may disrupt wildlife habitats and migration routes.

- The combustion of methane produces CO_2, a known greenhouse gas. Methane itself is a greenhouse gas too. Any methane that is lost or leaked in exploration, recovery, and production contributes to the overall problem.

PROBLEM 10-3
Methane is widely used for home heating. If electric power plants rely increasingly on methane, won't this put a strain on the supply, causing ever-more-severe problems with wintertime price volatility and spot shortages?

SOLUTION 10-3
This is an important issue, and it is one of the main arguments used by advocates for increased construction and deployment of nuclear, hydroelectric, solar, wind, geothermal, and even coal-fired power-generating facilities.

Onsite Combustion Generators

Small and medium-sized *combustion generators* are available for use in homes, businesses, hospitals, and government buildings. Some generators are also suitable for use by campers. For people living in remote areas, a combustion generator may be the primary, if not the only, source of electricity.

HOW THEY WORK

Figure 10-6 is a functional diagram of a typical combustion generator. The unit shown in this example provides 117 V AC, which is the usual output of portable generators used to power small appliances such as lamps and radios. Larger generators also supply 234 V AC for heavy appliances such as electric ranges and laundry machines. The engine can range in size from a few horsepower (comparable to the engine in a lawn mower) to hundreds of horsepower (comparable to the

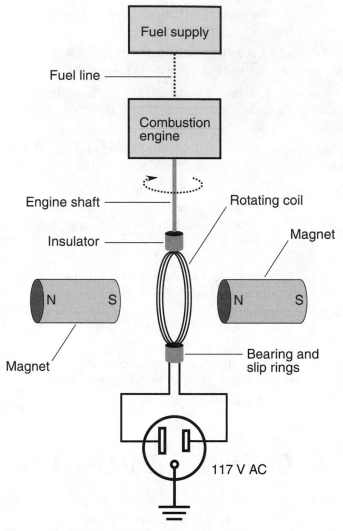

Figure 10-6 Simplified functional block diagram of a small- or medium-sized onsite combustion generator.

engines in trucks, tractors, and construction equipment). Most small generator engines burn gasoline. Larger ones use diesel fuel, propane, or methane.

In the AC generator, a coil of wire, attached to the shaft of the combustion engine, rotates inside a pair of powerful magnets. If a load is connected to this coil, an AC voltage appears across its end terminals as each point in the wire coil moves across *lines of flux* produced by the magnets, first in one direction and then in the other direction, over and over. (In an alternative arrangement, the magnetic poles revolve around the wire coil, which remains fixed.) The AC voltage that a generator can produce depends on the strength of the magnets, the number of turns in the wire coil, and the speed of rotation. The AC frequency depends only on the speed of rotation. In the United States, this speed is 3600 revolutions per minute (3600 rpm) or 60 revolutions per second (60 rps), so the output frequency is 60 cycles per second (60 Hz). In order to maintain a constant rotational speed for the generator under conditions of variable engine speed, mechanical regulating devices are used.

When a load is connected to the output of a generator, it becomes mechanically harder to turn the generator shaft, as compared to when there is nothing connected to the output. As the amount of electrical power demanded from a generator increases, so does the mechanical power required to drive it, and thus the amount of fuel consumed per unit time. The electrical power that comes out of a generator is always less than the mechanical power required to drive it. The lost energy occurs mainly as heat in the generator components. The *efficiency* of a generator is the ratio of the electrical power output to the mechanical driving power, both measured in the same units (such as watts or kilowatts), multiplied by 100 to get a percentage. No generator is 100-percent efficient, but a good one can come fairly close.

A typical small gasoline-powered generator provides from 1 kilowatt (1 kW) to 5 kW of electrical power. Such a generator, if not well designed, may cause problems when you try to run sensitive electronic equipment from it. However, a well-engineered generator, even the small gasoline-burning type, will work fine with computers and other sophisticated systems as long as it is kept in proper working order. Medium-sized diesel-, propane-, and methane-fueled generators can supply several tens of kilowatts, and can power an entire home, business, or agency. Large institutions typically have multiple generators. These machines, if properly operated and maintained, can operate all kinds of equipment, even the most sensitive and complex medical devices.

An onsite standby generator must be run only when the wiring in the building has been completely separated from the electric utility by means of a *double-pole, double-throw* (DPDT) *isolation switch*. Otherwise, *backfeed* can occur, endangering utility workers and damaging electrical system components. In the United States, the DPDT isolation arrangement is required by the *National Electric Code* (NEC).

ADVANTAGES OF ONSITE COMBUSTION GENERATORS

- A well-maintained and properly operated onsite generator can eliminate much, if not all, of the inconvenience associated with utility power blackouts.

- In a critical setting such as a hospital, onsite standby generators can save lives.

- When properly installed, operated, and maintained, a combustion generator can provide power any time it is needed, and for as long as it is needed. This is not in general true of alternative electric energy sources such as stand-alone photovoltaic (solar) systems or wind-driven systems.

- Well-designed and well-maintained combustion generators are rugged, reliable, and reasonably efficient.

- Combustion generator technology is adaptable. For example, generators fueled by petroleum diesel can be adapted to burn biodiesel. Methane-fueled generators can burn methane derived from biological process as well as the more familiar product of natural-gas or coal refining.

LIMITATIONS OF ONSITE COMBUSTION GENERATORS

- Combustion generators produce exhaust with the same pollutants that come from combustion type furnaces and motor vehicles.

- Onsite combustion generators can be dangerous. All instructions and electrical codes must be strictly followed.

- An onsite combustion generator must be provided with an uninterrupted supply of fuel for as long as its service is needed. In the case of gasoline, diesel fuel, and propane, this requires onsite storage tanks.

- Methane-powered generators will not work if the utility is interrupted. This can be the direct result of a catastrophe such as an earthquake. Methane utilities routinely shut off the gas during violent disasters to prevent flash fires and explosions in the event a line is ruptured.

PROBLEM 10-4

I'd like to build an energy-independent "get away from it all" house in the desert of northern Nevada. It's cold there in the winter, but the sun shines a lot, and the wind is a reliable resource as well. I'd like to use an onsite combustion generator to supplement a stand-alone solar-power system and a stand-alone wind-power system. Is that a workable idea?

SOLUTION 10-4

With all three of these power sources available, you should never have to worry about an electricity blackout! Be sure there's plenty of fuel available for the combustion generator in case there is a prolonged spell of cloudy, windless weather. Keep in mind that a comprehensive system of this sort, while offering independence from the electric utility, will cost you a lot of money up front. In addition, fuel and maintenance will impose ongoing expenses. You should size each system to ensure that you will always have the amount of electrical power you need. The services of a competent engineer will be required if the systems share any wiring. Otherwise you might have trouble with electrical conflicts (known as *bucking*) between different systems. This condition, which can damage system components as well as equipment connected to the electrical wiring, occurs when the waves from two or more AC sources are not kept in phase (perfect lock-step) with each other.

Quiz

This is an "open book" quiz. You may refer to the text in this chapter. A good score is eight correct. Answers are in the back of the book.

1. Methane can be derived from
   a. carbon dioxide.
   b. gasoline.
   c. coal.
   d. nitrous oxides.

2. If the rotational speed of the shaft connected to the coil in an electric generator changes, which of the following will occur?
   a. The AC frequency will change.
   b. The load resistance will decrease.
   c. The power drawn by the load will decrease.
   d. All of the above

3. Which of the following, a, b, or c, if any, is an advantage of coal trains over extremely long high-voltage transmission lines in getting coal-generated electricity to end users?
   a. Coal trains cost less to operate than extremely long high-voltage power lines would cost to build and maintain.
   b. Coal trains cause no serious trouble of any kind, while high-voltage power lines generate dangerous NO_x, SO_x, and CO pollution.
   c. Coal trains require less fuel per kilometer of delivery than extremely long high-tension lines would require for the transport of energy.
   d. There aren't any advantages to coal trains over extremely long high-voltage transmission lines.

4. Bucking and backfeed among two sources of AC electricity can be prevented by

    a. making sure the wires are connected with the same polarity in each system.

    b. using three-wire, grounded electrical wiring in all systems.

    c. keeping the wiring for the two systems electrically isolated.

    d. never running more than one source of electric power at a time.

5. An advanced steam-turbine coal-fired power plant can be adapted to burn other fuels. Which of the following is one of these alternatives?

    a. Methane

    b. Carbon monoxide

    c. Nitrous oxide

    d. Sulfur dioxide

6. In a combined-cycle electric-generation system, the mechanical torque that drives the generators comes directly from

    a. gas and steam turbines.

    b. step-up transformers.

    c. water boilers.

    d. oil or methane combustion.

7. If the voltage in a power transmission line could be increased by a factor of 50, assuming constant load and constant wire resistance, then the *current* in the line would

    a. stay the same.

    b. increase by a factor of 50.

    c. decrease by a factor of 50.

    d. do none of the above.

8. If the voltage in a power transmission line could be increased by a factor of 50, assuming constant load, then the *power loss* in the line would

    a. stay the same.

    b. increase by a factor of 50.

    c. decrease by a factor of 50.

    d. do none of the above.

9. The health hazard produced by the electricity in a high-tension transmission line can be minimized by

    a. maximizing the frequency of the AC carried by the line.

    b. minimizing the current carried by the line.

    c. maximizing the power consumed by the load.

    d. minimizing the resistance of the load.

10. A combustion-turbine type oil-fired power plant contains all of the following components except

    a. a transformer.

    b. a compressor.

    c. a gas turbine.

    d. a pulverizer.

CHAPTER 11

Electricity from Water and Wind

Moving water and air are among the most natural forms of alternative energy. Nothing on this planet is truly infinite in supply, but the energy available from these sources, in practical terms, comes close to that ideal. Only solar energy is more nearly eternal.

Large- and Medium-Scale Hydropower

A boom in hydroelectric power plant construction took place in the United States around the beginning of the 20th century. As time passed, other types of power plants came online, including those fueled by coal, oil, methane, and nuclear fission. In America today, only a small fraction of electricity is generated by hydroelectric power plants, which can exist in three different forms. The best hydroelectric technology for a given location depends on the nature of the terrain, the present and anticipated future need for electricity, and the effects of the facility on plants and animals, water quality, agriculture, and overall quality of life.

IMPOUNDMENT

An *impoundment hydroelectric power plant* consists of a dam and reservoir. This type of facility works best in hilly or mountainous terrain where high dams can be built and deep reservoirs can be maintained.

The potential energy available in a reservoir depends on the mass of water contained in it, as well as on the overall depth of the water. The potential energy in a specific parcel of water is expressed in *newton-meters* (N · m). The *newton* is the standard unit of force, equivalent to a meter per second squared (1 m/s^2). This is the product of the mass of the parcel in kilograms, the acceleration of gravity in meters per second squared (about 9.8 m/s^2), and the elevation of the parcel in meters (the vertical distance it falls as its energy is harnessed). The equivalent kinetic-energy unit is the *joule* (J), which is in effect equal to a watt-second (W · s).

Figure 11-1 is a simplified functional diagram of an impoundment facility. The water from the reservoir passes through a large pipe called a *penstock*, and then through one or more water turbines that drive one or more electric generators. The

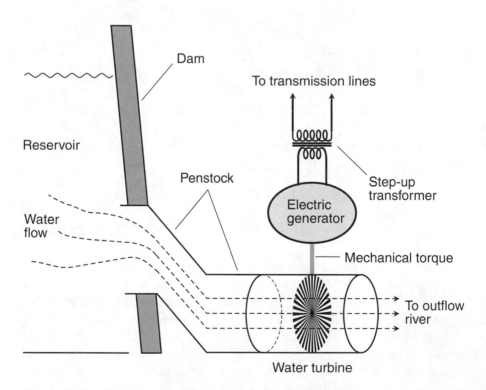

Figure 11-1   Simplified functional diagram of a large generating system that derives its energy from water impoundment.

output of each generator is stepped up in voltage and then sent to the transmission line for distribution.

DIVERSION

In a *diversion hydroelectric power plant*, a portion of the river is channeled through a canal or pipeline, and the current through this medium is used to drive a turbine. A dam is not required. This type of system is best suited for locations where a river drops considerably per unit of horizontal distance. The ideal location is near a natural waterfall or rapids. The chief advantage of a diversion system is the fact that, lacking a dam, it has far less impact on the environment than an impoundment facility.

Small- and medium-scale diversion systems can be used next to mountain streams or other fast-moving, small rivers for the purpose of providing energy to individual homes or subdivisions. This type of system can be adapted to electric resistance heating systems, as discussed in Chapter 3. However, a small- or medium-scale diversion system will not work if the stream or river dries up, or if it freezes completely from the surface to the bottom.

PUMPED-STORAGE

A *pumped-storage hydroelectric power plant* has two or more reservoirs at different elevations. When there is little demand for electricity among the consumers in the region served by the facility, the excess available power is used to pump water from the lower reservoir into the upper one(s). When demand increases, the potential energy stored in the upper reservoir(s) is released. Water is let out of the upper reservoir(s) in a controlled manner, passing through penstocks and turbines to generate electricity.

Pumped-storage systems require dams to hold the water in the reservoirs. These dams are generally smaller than those used in large impoundment facilities. Pumped-storage power plants can be found in regions where the terrain is hilly or gently rolling. But there must be a significant difference in average elevation between the reservoirs.

ADVANTAGES OF HYDROELECTRIC POWER PLANTS

- Hydroelectric power plants generate no CO_2, CO, NO_x, SO_x, particulates, ground contamination, or waste products. Some heat is imparted to the stream or river water as a result of friction with the turbine components, but this is not always significant.

- Water is a renewable source of energy, as long as the stream or river does not dry up. The hydrologic cycle replenishes the source of potential energy in the form of rainfall, snowfall, and runoff.

- Hydroelectric power plant output can be controlled at will by changing the volume of water flow per unit time.

- The reservoirs created by an impoundment or pumped-storage power plant can be used for recreational purposes, and may also provide dramatic scenery.

- The reservoirs created by impoundment are generally clean, because impurities precipitate to the bottom. They can often be used as sources of water for drinking, bathing, washing, or irrigation.

LIMITATIONS OF HYDROELECTRIC POWER PLANTS

- Large reservoirs flood land that could be used for something else. Whole towns have been sacrificed to reservoirs, causing displacement, resentment, and economic hardship.

- If a dam fails in a large impoundment facility, a catastrophic flood is almost certain to occur downstream.

- Hydroelectric power plants are not practical in regions where the terrain is flat.

- A prolonged drought can adversely impact, or even cut off, the energy production capacity of a hydroelectric power plant.

- In impoundment and pumped-storage power plants, the water level in the reservoir(s) varies considerably. One cannot expect to build a "beach house" directly on a reservoir!

- A dam can cause low dissolved-oxygen levels in the reservoir, because it brings the normal river flow to a nearly complete halt. This can kill fish and affect the nature of plant life in and around the reservoir.

- Dams can interfere with fish spawning. This problem can be mitigated by the use of *fish ladders*, *fish elevators*, or trapping and hauling the fish. However, such measures add to the cost of system construction or operation.

PROBLEM 11-1

In light of the problems with fossil fuels and the dangers of nuclear fission, why can't more hydroelectric power plants be built? There are plenty of rivers on all the continents of the world. Shouldn't we build as many hydroelectric facilities as we can?

SOLUTION 11-1
Most good sites for large hydroelectric power plants have been exploited already. There is a practical limit to the number of dams and reservoirs that can be placed along any river. Whenever energy is taken from a river by a hydroelectric facility, that energy is not available for use in any form further downstream. If too many hydroelectric power plants are placed along a river, energy availability and economic conflicts occur.

Small-Scale Hydropower

In *small-scale hydropower systems*, diversion technology is most commonly used, although impoundment is sometimes done to a limited extent. A water turbine designed for home or small business use, installed in a fast-moving stream or small river with sufficient vertical drop, can produce 20 kW of electricity, more than enough for a typical household under conditions of peak demand.

HOW IT WORKS

A small-scale hydropower system can be configured in three ways: stand-alone, interactive with batteries, and interactive without batteries. Interactive systems are also known as *intertie systems* or *grid-intertie systems*.

A stand-alone system employs banks of rechargeable batteries to store some or all of the electric energy supplied by the water turbine. The batteries supplement the power from the turbine, and can provide all the electricity if the turbine ceases to produce power. An interactive system with batteries uses the electric utility, rather than the turbine, to keep the batteries charged. In an interactive system without batteries, excess energy is sold to the utility during times of minimum demand, and energy is bought from the utility during times of heavy demand. Some states offer good buyback deals with utility companies. Some states do not.

Further details about how these three types of systems operate are provided in the section "Small-Scale Wind Power" later in this chapter. The principles are basically the same for hydropower, wind power, and solar power.

ADVANTAGES OF SMALL-SCALE HYDROPOWER

- Small-scale hydropower can reduce or eliminate dependence on conventional electric utilities.

- Water flow is continuous, as long as the stream is large and fast enough. It's more reliable than wind or solar energy sources, as long as the stream doesn't dry up or freeze solid.

- Small-scale hydropower plants are virtually nonpolluting. A small amount of heat is imparted to the stream water as a result of friction with the water turbine components, but this is rarely significant.

- The electricity produced by a water turbine can be used for supplemental home heating or evaporative cooling.

LIMITATIONS OF SMALL-SCALE HYDROPOWER

- Few people live on properties with streams running through that provide enough flow to provide hydroelectric power.

- A small stream may periodically freeze or dry up, shutting a small-scale hydropower system down.

- A water turbine requires considerable water mass, along with a significant vertical drop, in order to provide enough power to heat a home. This may necessitate the installation of a small dam or artificial waterfall, which could give rise to environmental and regulatory issues.

- The up-front cost of a small-scale hydropower system is considerable. It takes a long time to pay for itself, and the resulting economic benefit may be outstripped by the initial cost.

PROBLEM 11-2
I'd like to install a stand-alone small-scale hydropower system for my home. I live on a ranch. A fairly good-sized stream runs through my property. An engineer has checked everything out. The vertical drop is sufficient, and there is more than enough water flow all year round. I'll need to build a small dam and back up some water to form a pond. This is all right with the local, state, and federal officials. But I'm concerned about how this system will affect wildlife.

SOLUTION 11-2
The wildlife-impact question is best answered by getting a wide variety of opinions. Naturalists from a nearby college or university should be able to offer some insight. A pond can be expected to attract birds, fish, and other wildlife (some wanted, some not!).

Tidal-Electric Power

Ocean tides are caused by the gravitational fields of the moon and the sun, in conjunction with the rotation of the earth on its axis. In a simplistic sense, tides are waves with an extremely long *period* (time for the completion of a cycle), with two *crests* (high points) and two *troughs* (low points) each day in most locations.

THE TIDAL BARRAGE

A *tidal barrage* resembles a small dam with *sluice gates* that can be opened or closed, allowing water to flow between bodies of water having different elevations. This flow operates a water turbine that is mechanically coupled to an electric generator. The principle of operation is shown in Figure 11-2.

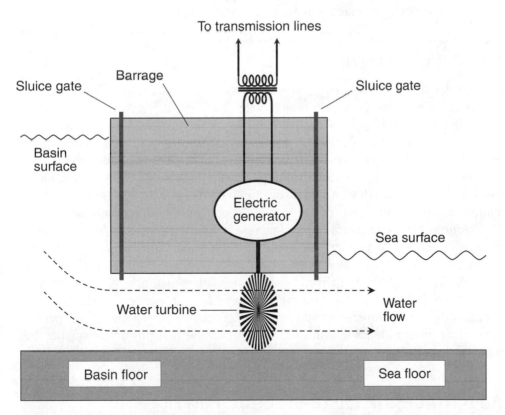

Figure 11-2 Simplified functional diagram of a large generating system that derives its energy from a tidal barrage.

As the tide comes in, the basin is allowed to fill through a large channel (not shown) until the tide reaches its highest point. The sluice gates are closed during this time, and the turbine does not operate. When the tide is at its peak, the elevation of the surface of the basin is the same as the level of the sea surface outside the basin. Then the tide begins to fall, and the basin acts as a reservoir. The sluice gates are opened so water flows through the turbine, powering the generator. When the tide reaches its lowest point, water continues to flow through the turbine. But shortly after low tide, as the ocean level begins to rise again, the surface elevations of the basin and the ocean become equal. Then the sluice gates are closed, the turbine stops, and the basin is allowed to fill as the tide comes in.

An alternative system takes advantage of the incoming tide as well as the outgoing tide. Two basins are used, one constantly maintained at a level above that of the ocean and the other constantly maintained at a level below that of the ocean. The ideal tidal power plant of this sort has two systems like that shown in Figure 11-2 operating back-to-back with turbines facing in opposite directions. Although this can theoretically double the amount of energy produced over time, it also costs approximately twice as much to build as a single-ended system.

THE TIDAL TURBINE

As the tides flow in and out in the vicinity of an irregular shoreline, currents are set up in the water. *Littoral currents* flow parallel to the shoreline. *Rip currents* flow in large eddies near the shore. Tide-like currents can be produced by offshore storm systems. There are also large-scale, persistent ocean currents in various parts of the world. A good example is the *gulf stream* that flows eastward out of the Gulf of Mexico, around the tip of the Florida peninsula, and across the Atlantic Ocean toward the British Isles. Another example is the *Alaska current* that flows southeastward off the coasts of California, Oregon, and Washington State.

A *tidal turbine* bears a strong physical resemblance to a *wind turbine* (discussed later in this chapter). The turbine, and its associated support, are anchored to the ocean floor. Multiple turbines create a *tidal stream farm*. Each turbine is connected to an electric generator. The entire system is beneath the surface, so it is not visible from above. Ocean currents travel more slowly than atmospheric winds, but water is hundreds of times more dense than air, so it produces much greater force per unit area on the turbine blades. For this reason, tidal turbines are physically smaller than wind turbines.

ADVANTAGES OF TIDAL-ELECTRIC POWER

- The tides are a renewable, reliable, and predictable resource.

- In a location with a significant difference between high and low tide, the ebb and flow can be harnessed to produce electrical power on a consistent basis.

- Tidal-electric power systems, like hydroelectric systems, generate no CO_2, CO, NO_x, SO_x, particulates, ground contamination, or waste products. Some heat is imparted to the ocean as a result of friction with turbine components, but this is rarely significant.

- Tidal-electric power systems seem exotic to some people. The existence of this type of facility in a given area can therefore be used to promote tourism, bringing in revenue.

- A tidal barrage can serve as a bridge for a roadway or railway across a bay or estuary.

- Maintenance of a tidal barrage is not difficult. The turbines last upwards of 30 years, and the barrage itself is inherently simple. However, the initial cost of installation is high.

- Tidal turbines are entirely beneath the surface. If installed in deep enough water, they present no obstacle or hazard to marine transportation.

LIMITATIONS OF TIDAL-ELECTRIC POWER

- The installation of a tidal barrage is an expensive undertaking. Once installed, however, maintenance is relatively easy.

- Tidal turbines can be difficult to install, because the best sites for tidal currents are often in treacherous waters near rugged coastlines.

- Tidal-electric power plants can have a negative effect on marine life. Large fish, turtles, and marine mammals can be killed by the turbines, especially in tidal barrage systems. A large "catch" of this sort can also damage a turbine.

- A tidal barrage creates a sort of reservoir out of a bay or estuary, modifying its characteristics. This affects the *turbidity* (cloudiness) of the water and the extent of *sedimentation* (the settling of solid particles to the bottom).

- If not carefully designed and operated, a tidal barrage can cause localized flooding.

PROBLEM 11-3

Why are two systems of the sort shown in Figure 11-2, operating back-to-back, necessary in order to generate tidal power all the time? Can't a single-basin system be designed that takes advantage of the incoming tide as well as the outgoing tide, thereby generating power continuously?

SOLUTION 11-3

A single-basin barrage system can be built that provides power *almost* all the time, but there are technical problems. Imagine a system with two sluices (call them the *inflow sluice* and the *outflow sluice*), each containing a turbine. Suppose the system is designed so the elevation of the basin lags the elevation of the sea by one-quarter of a complete tidal cycle. When the level of the basin is higher than that of the sea, water passes through the outflow sluice, and the inflow sluice is closed. When the level of the basin is lower than that of the sea, water passes through the inflow sluice, and the outflow sluice is closed. Therefore, the basin and the sea have tides of equal magnitude but with different peak times, as shown in Figure 11-3. The available power at any given moment in time depends on the difference in elevation between the sea and the basin, represented by the vertical distance between equal-time points on the curves. (At points where the curves intersect, no power is available because the levels of the sea and the basin are the same.) Unfortunately, this kind of system suffers from variability in power output. Moreover, it cannot provide any more total energy per unit tide cycle than a single-ended system. The most efficient and cost-effective way to get continuous energy from the tides is to employ two or more back-to-back systems similar to that diagrammed in Figure 11-2, with independent basins and with their flow cycles timed so at least one of the systems always provides power.

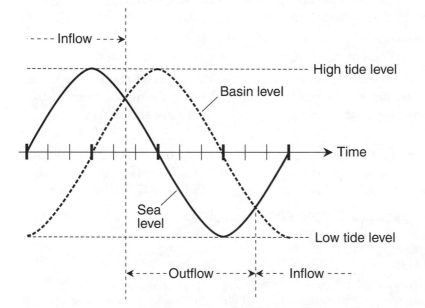

Figure 11-3   A tidal barrage system in which the elevation cycle of the basin lags the elevation cycle of the sea by one-quarter of a tide cycle.

Wave-Electric Power

On large bodies of water, where the *fetch* is long (the wind blows over the surface for a great distance), friction between the air and the water surface causes *ripples* that grow into *wavelets*, then into *waves*, and ultimately into *swells*. Swells typically measure from 1 m to 3 m from crest to trough in the ocean near shore. Some swells, such as those generated by large low-pressure systems and hurricanes, can exceed 25 m (80 ft) from crest to trough. Ocean swells contain the capacity to generate usable power. Normally this power is dissipated when the swells reach the shore in the form of *breakers*. A *wave-electric generator* taps the power of ocean swells and converts it into electricity.

HOW IT WORKS

Have you ever been at a swimming pool where artificial waves were made? This can be done by pumping air in and out of a partially submerged chamber. If the chamber is airtight and watertight at all points that stay above the water surface (except for a hole where a pump is connected) but water-transparent at all points that stay below the surface, the water in the chamber cyclically falls and rises as air is pumped in and out of the hole. The wave maker converts the mechanical power from the air pump into power that propagates outward through the water.

Now imagine this process in reverse. Suppose some type of chamber is placed a short distance offshore in a turbulent sea. This chamber, like the swimming-pool wave maker, has a hole above the surface. If the chamber is attached firmly to the bottom so it cannot move up and down, water swells pass through it and cause the level of the water inside to rise and fall. The surface inside stays essentially flat as its level fluctuates, so air is pushed out of, and then drawn back into, the hole. The hole contains an *air turbine* similar to the gas turbine in a jet engine. The turbine rotates as the air is pumped in and out. The resulting mechanical torque drives an electric generator. The output of the generator passes through a *power converter* that changes its electricity to 60-Hz AC (or 50-Hz AC in Europe) for distribution over electric transmission lines. Figure 11-4 is a functional diagram of such a system.

ADVANTAGES OF WAVE-ELECTRIC POWER

- The turbulence of the world's oceans is a renewable resource.
- The conversion of wave power to electricity does not generate CO_2, CO, NO_x, SO_x, particulates, ground contamination, or waste products.

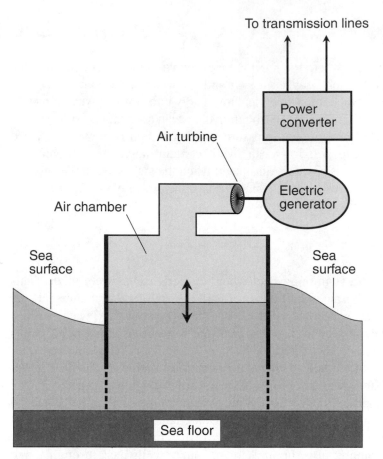

Figure 11-4   Simplified functional diagram of a wave-electric power-generating system.

- A wave-electric generator is not particularly expensive to install or maintain, as long as it is engineered to withstand storms (without wasteful overengineering).

- Large "wave-electric power farms" can produce great quantities of usable electricity.

- Wave-electric generators have a low profile. Even when observed, they blend in fairly well with the scenery. (However, this can also be a problem; note the last limitation on the next page.)

- Wave-electric generators, if properly designed, do not have a significant adverse effect on marine life.

LIMITATIONS OF WAVE-ELECTRIC POWER

- When the ocean surface is calm or nearly calm, a wave-electric generator will not produce usable output.

- Wave-electric generators must be sited carefully to minimize the effects of the noise they produce, but they must nevertheless be located where the energy from swells is available in sufficient amounts.

- A "hundred-year storm" may destroy a wave-electric generator unless it is overengineered to the extent that its cost does not justify its use.

- Wave-electric generators, because of their low profile, may present a hazard to marine navigation unless their presence is made clear on maps. Buoys or other markers may be necessary.

PROBLEM 11-4

Are there types of wave-electric generators besides the one diagrammed in Figure 11-4?

SOLUTION 11-4

Yes. In one design, a string of floats, connected together by hinges, is placed on the water surface. The entire assembly undulates, and the resulting mechanical torque at the hinges can drive electric generators. Another design employs water turbines placed on the sea floor near a shoreline where strong *undertows* occur. An undertow is a current that flows away from shore after water hurled onshore by a breaker recedes down the sloping bottom. Other designs can take advantage of the pressure caused by breakers as they interact with seawalls or other obstructions placed at the shoreline.

Large-Scale Wind Power

The wind is one of the oldest sources of energy harnessed by humankind. *Windmills* have been used for centuries to pump water and to mill grain (that's where the "mill" in "windmill" comes from). Nowadays, wind power is making a comeback as problems with conventional energy sources increase.

WIND SPEED, OPERATING RANGE, AND POWER

Wind speed can be specified in *meters per second* (m/s), *kilometers per hour* (km/h), *statute miles per hour* (mi/hr), or *nautical miles per hour* (nmi/h), also known as *knots* (kt). The unit most often used by weather forecasters and other professionals is the knot, which is equivalent to approximately 1.852 km/h or 1.151 mi/h. The unit preferred by mass-media news and weather broadcasters in the United States is the statute mile per hour, which is equivalent to approximately 1.609 km/h or 0.8690 kt.

Manufacturers of *wind turbines*, especially in Europe, quantify wind speed in meters per second. If you are told the wind speed in meters per second, multiply by 2.237 to determine the wind speed in miles per hour, or by 1.943 to determine the speed in knots. If you are given the wind speed in miles per hour, divide by 2.237 to determine the speed in meters per second. If you are given the wind speed in knots, divide by 1.943 to determine the speed in meters per second.

Most large wind turbines are designed to operate at wind speeds ranging from 3 to 4 m/s up to 20 to 25 m/s. That corresponds to a minimum or *cut-in* wind speed of 7 to 9 mi/h, a breeze that lightly fans the face, and a maximum or *cut-out* wind speed of 45 to 56 mi/h, a gale against which it is difficult to walk. Within this operating range, a large wind turbine can generate anywhere from a few hundred kilowatts (kW) up to several megawatts (MW) of usable electric power, depending on the blade length, the wind speed, and the size of the generator.

The *capacity factor* of wind energy—the proportion of time the resource can be exploited to produce usable output—is approximately 25 to 40 percent, depending on the geographic location and on the design of the turbine. The best sites for wind turbines are often far from population centers, necessitating the use of long transmission lines to get electricity to end users.

DESIGN CONSIDERATIONS

Large wind turbines require tall, strong towers for support. A typical wind turbine tower measures between 50 m (165 ft) and 80 m (260 ft) high, and is anchored in a mass of concrete. Some towers are guyed for enhanced high-wind survival.

The blades in a large wind turbine range from approximately 25 m (80 ft) to 40 m (130 ft) in length. Most large wind turbines have three blades. Some designs have two blades, and a few variants have four or more. The *rotational diameter* is twice the blade length. The blade bearing, electric generator, and generator cooling apparatus are contained in a housing called a *nacelle*. The system is designed to spin at a constant *angular speed* of approximately 20 revolutions per minute (rpm) for the entire workable range of wind speeds. Changes in wind speed within the workable range cause variations in the maximum deliverable output power, but not in the rate of blade rotation. A gear box translates the angular speed of the blades into the proper angular shaft speed for a generator to produce AC at a constant frequency of 60 Hz (in the United States) or 50 Hz (in Europe and some other parts of the world).

In order to function properly, a large wind turbine must be oriented so its blades rotate on an axis that points into the wind. This means that the plane defined by the rotating blades must be perpendicular to the wind direction. To orient itself, the entire assembly (blades and nacelle) can swivel through a full 360° horizontal circle on a *turntable*. The wind direction and wind speed are detected by a *wind vane* and an *anemometer* similar to the instruments used by meteorologists for the same purposes. If the wind becomes too strong, a *fail-safe braking system* stops the blades and locks them in place, and the turntable rotates approximately 90° so the assembly experiences the smallest possible wind load.

In most large wind turbines, the rotor blades are on the windward side of the nacelle, as shown in Figure 11-5A. This is called *upwind design*. In some wind turbines, the rotor blades are on the leeward side of the nacelle (see Figure 11-5B). This is known as *downwind design*. Debate has taken place concerning which design produces more electrical energy in the long term for a given amount of money spent. Proponents of the upwind design emphasize the fact that the technology is proven, and that significant improvements in efficiency and durability have taken place since the turn of the millennium. Those who favor downwind design argue that the blades can be designed to "fold" or "cone" away from the wind as they

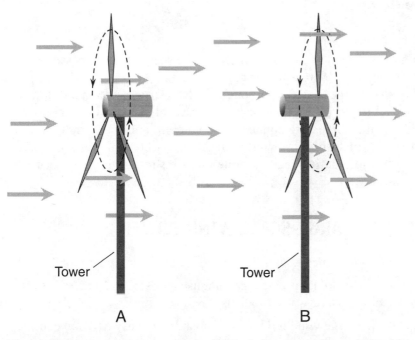

Figure 11-5 At A, a large upwind turbine. At B, a large downwind turbine. Gray arrows represent the wind.

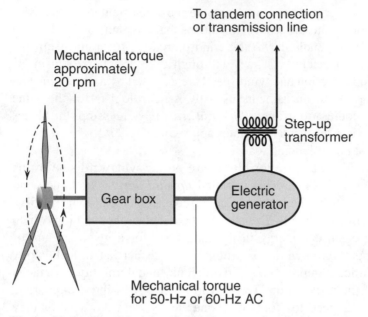

To tandem connection
or transmission line

Mechanical torque
approximately
20 rpm

Step-up
transformer

Gear box

Electric
generator

Mechanical torque
for 50-Hz or 60-Hz AC

Figure 11-6 Simplified functional diagram of a wind-power generator using a large turbine and transformer.

rotate, allowing the turbine to function at higher wind speeds than is possible with an upwind turbine, thereby increasing the capacity factor.

Figure 11-6 shows the basic components of a large-scale wind turbine connected to the utility grid. In some large-scale wind-power systems, groups of up to a half dozen turbines are connected in tandem, and the combination is connected to the grid. In the biggest wind farms, dozens of turbines are connected in tandem, and the combination feeds a cable that runs to a switchyard or high-tension transmission line.

ADVANTAGES OF LARGE-SCALE WIND POWER

- Wind is a renewable energy source, and the supply is practically unlimited.
- Wind-power plants do not produce greenhouse gases, CO, NO_x, SO_x, particulate pollutants, or waste products.
- Once installed, a large wind turbine is comparatively easy and inexpensive to maintain.
- Wind-power plants can reduce dependency on fossil fuels, hydropower, and nuclear fission reactors for electric generation.

- Large wind turbines can be dispersed over wide regions. This distributes the energy source and should help in the quest for a *fault-tolerant* utility grid (a system not vulnerable to catastrophic failure or sabotage).
- Wind power can be used to supplement other modes of electric power generation. This increases the diversity of a nation's electrical system.
- Large wind turbines can be placed offshore over large lakes or over the ocean, as well as on land.
- Even the largest wind turbines have small footprints, and can thus share land resources with other operations such as farming and cattle ranching.

LIMITATIONS OF LARGE-SCALE WIND POWER

- The wind is an intermittent source of energy. The capacity factor is lower than that of most other energy sources.
- A large wind turbine can be damaged or destroyed by a severe thunderstorm, hurricane, or ice storm.
- Some people do not like the physical appearance of large wind turbines.
- Wind turbines make some noise. However, at a reasonable distance from the tower, blade and turbine noise is rarely much louder than the wind itself.
- Large wind turbines may occasionally injure or kill birds. This problem can be mitigated by judicious choice of location, and by not placing multiple turbines wall-to-wall (in close proximity with a common plane of blade rotation).
- Wind power cannot, by itself, totally satisfy the electrical needs of a city, state, or nation. It is at best a supplemental source, used in conjunction with fossil fuels, nuclear fission, and hydropower.
- Locations with consistent usable winds are often far from population centers, requiring the use of long transmission lines.

PROBLEM 11-5

Why hasn't wind power been exploited to a greater extent? If wind turbines are distributed throughout the windy regions of the United States, then there ought to be good winds blowing at some of the sites, no matter what the time. Wouldn't this provide a constant supply of electricity if all the turbines were tied together into a single grid?

SOLUTION 11-5

In theory, the answer to this question is "Yes." At any given time, there are plenty of locations in the United States where the wind is blowing at optimum speeds for the operation of wind turbines. The problem is finding a way to efficiently get the electricity from generating points to end users. Invariably, some end users are too far from operational generators to allow efficient transfer of the electricity to them. Technology does not yet exist to store wind-derived energy in sufficient quantity to serve cities and towns on a continuous basis.

Small-Scale Wind Power

The term *small-scale* applies to wind turbines that can generate as much as 20 kW of electricity under ideal conditions, enough to power most households. Like their larger counterparts, small-scale wind turbines generate power on an intermittent basis. In order to obtain a continuous supply of electricity with a small-scale wind-power plant, it is necessary to use storage batteries or an interconnection to the electric utility, or both.

HOW IT WORKS

Most small-scale wind turbines are steered by a *wind vane* attached to the nacelle, rather than by a powered turntable. The vane works in the same way as an old-fashioned weather vane. When the wind is strong enough to operate the turbine, the vane orients itself to point away from the wind, and the whole turbine resembles a miniature single-engine propeller airplane without the wings. Under normal operating conditions, the plane defined by the blade rotation is broadside to the wind.

In a small-scale wind-power system, the speed of the blade rotation varies with the wind speed. This results in variable-frequency AC from the generator inside the nacelle. This generator is similar to the *alternator* in a car or truck (some manufacturers actually call it an alternator). The AC from the generator is converted to DC by a *rectifier* circuit, and the DC is used to charge a set of storage batteries. The electricity for household appliances is derived from these batteries either directly, in which case special DC appliances must be used, or by means of a *power inverter* that converts low-voltage DC from the batteries to 117 V AC at 60 Hz (in the United States) or 50 Hz (in Europe and some other parts of the world).

When the wind speed exceeds a specified level, a small-scale wind turbine turns sideways to the wind to some extent. The plane defined by the blades is normally perpendicular to the axis of the vane. However, in a strong wind, the plane of the blades changes, so it is no longer perpendicular to the vane axis. This reduces the wind load on the blades while still allowing the turbine to function. As the wind speed grows stronger yet, the angle between the plane of the blades and the vane axis decreases until, at a certain speed, it becomes zero. Then the blades rotate in a plane that contains the axis of wind flow. The variation in the angle between the plane of the blades and the wind direction is called *furling*. It can be done in the horizontal plane (so the blades swing, or *yaw*, toward the left or right) or in the vertical plane (so the blades tilt upwards).

Varying the *blade pitch* is another means by which a small-scale wind turbine can regulate its wind load. When the blade pitch is small (the plane of each blade's surface is nearly the same as the plane defined by the blades), the wind produces less force on the system, and consequently less power, than when the blade pitch is large (the plane of each blade's surface differs greatly from the plane defined by the blades). This scheme is similar to that used in propeller aircraft, but it works in the opposite sense. At low wind speeds, the blade pitch is maximum. As the wind speed increases, the blade pitch decreases. Ultimately, if the wind speed becomes great enough, the blade pitch becomes zero. In extremely high winds the blades can turn to zero pitch, furl completely, and lock in place. This reduces the load on the blades as much as possible, minimizing the risk of structural damage. It also shuts down the wind turbine.

STAND-ALONE SYSTEM

A *stand-alone small-scale wind-power system* employs rechargeable batteries to store the electric energy supplied by the rectified output of the generator. The batteries provide power to an inverter that produces a good AC wave at 117 V. In some cases the battery power is used directly, but this necessitates the use of home appliances designed for low-voltage DC. Figure 11-7 is a functional block diagram of a stand-alone small-scale wind-power system that can provide 117 V AC.

The use of batteries allows the system to produce usable power even if there is not enough, or too much, wind for the turbine to operate. A stand-alone system offers independence from the utility companies. However, a blackout will occur if the system goes down for so long that the batteries discharge and there is no backup power source. This type of system is normally designed for a single home. Battery technology is not practical for large-scale wind-power systems.

INTERACTIVE SYSTEM WITH BATTERIES

An *interactive small-scale wind-power system with batteries* is similar to a stand-alone system, but with one significant addition. If there is a prolonged spell in which wind conditions are unfavorable for turbine operation, the electric utility can take over to keep the batteries charged and prevent a blackout. A switch, along with a battery-charge detection circuit, connects the batteries to the utility

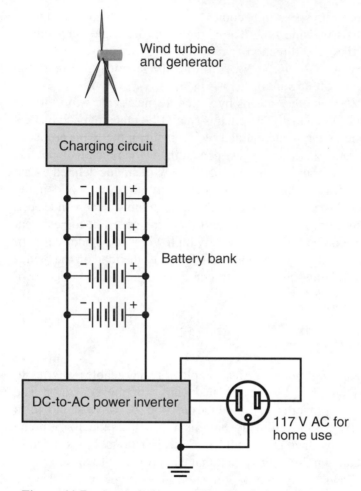

Figure 11-7   A stand-alone small-scale wind-power system.

through a charger if no power is coming from the turbine. When wind conditions become favorable and the turbine supplies power again, the switch disconnects the batteries from the utility charger and reconnects them to the turbine generator and rectifier.

Most interactive small-scale wind-power systems with batteries never sell any power to the electric utility, even if the wind turbine generates an excess. Power only flows one way, from the electric power line to the batteries through a charging circuit and switch, and even that happens only when the batteries require charging and the wind turbine does not provide enough power to charge them. Figure 11-8 is a functional block diagram of this type of wind-power system.

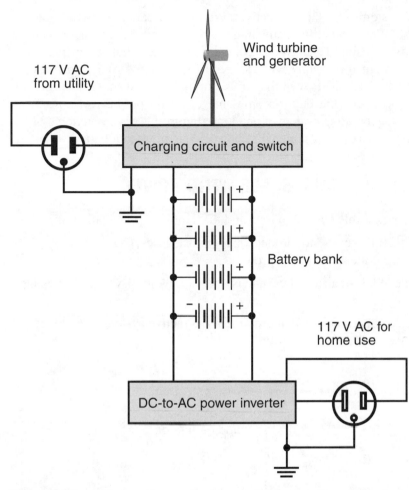

Figure 11-8   An interactive small-scale wind-power system with batteries.

INTERACTIVE SYSTEM WITHOUT BATTERIES

An *interactive small-scale wind-power system without batteries* also operates in conjunction with the utility companies. Energy is sold to the companies during times of minimum demand, and is bought back from the companies at during times of heavy demand. The advantage of this system is that you can keep using electricity (by buying it directly from the utilities) if wind conditions become unfavorable for a prolonged period. Another advantage is that, because no batteries are used, this type of system can be larger, in terms of peak power delivering capability, than a stand-alone arrangement or an interactive system with batteries.

This type of system, like the interactive system with batteries, is designed to function with the help of the utility companies, and does not offer the independence that a purist might desire. This isn't a technical drawback, but it can present a philosophical problem for anyone who desires to get totally off the grid. Some states offer good buyback deals with the utility companies, and some states do not. It's a good idea to check the buyback laws in your state before investing in an interactive electric generating system of any kind. Figure 11-9 is a functional block diagram of an interactive small-scale wind-power system without batteries.

ADVANTAGES OF SMALL-SCALE WIND POWER

- Wind is a renewable energy source, and the supply is practically unlimited.
- Small wind turbines do not produce greenhouse gases, CO, NO_x, SO_x, or particulate pollutants, or waste products.
- Once installed, a small wind turbine is comparatively easy and inexpensive to maintain.
- Small-scale wind power, when used in an interactive system, can reduce dependence on the electric utility.

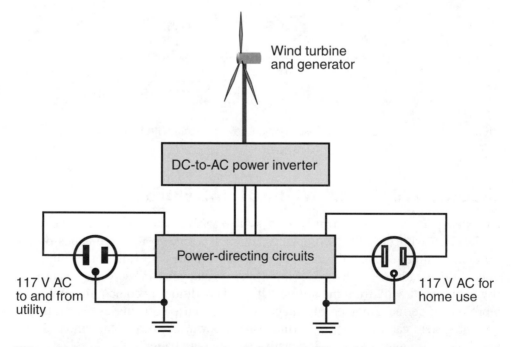

Figure 11-9　An interactive small-scale wind-power system without batteries.

- When used in conjunction with other alternative sources, small-scale wind power can offer complete independence from the utility.

- A small-scale wind-power system with batteries can provide electricity in the event of a short-term utility blackout.

- Small-scale wind power can be used for supplemental home heating or cooling.

LIMITATIONS OF SMALL-SCALE WIND POWER

- The wind is an intermittent source of energy. Even the best stand-alone small-scale wind-power system can provide only a small amount of electricity on a continuous basis.

- Small-scale wind turbines will not work properly if the wind is too strong.

- A small wind turbine can be wrecked by a powerful thunderstorm, hurricane, or ice storm.

- A small-scale wind-power system will take a long time to pay for itself, and in fact may never.

- Neighbors may dislike having a wind turbine nearby. (This is the "not in my back yard," or NIMBY, phenomenon.)

- Small-scale wind turbines can create noticeable noise. They rotate at higher speeds than large turbines, and are placed closer to the surface.

PROBLEM 11-6
What factors affect the amount of power that can actually be derived from the wind?

SOLUTION 11-6
The available power P, in watts, available from the wind depends on the air density, the area defined by the turbine blades as they rotate, and the wind speed. If d is the air density in kilograms per cubic meter (a value that varies somewhat with elevation, temperature, and barometric pressure), A is the area in square meters swept out by the blades in a plane perpendicular to the wind direction, and v is the wind speed in meters per second, then:

$$P = dAv^3/2$$

In theory, this means that the available power from a wind turbine varies in direct proportion to the air density, in direct proportion to the area swept out by the blades, and in direct proportion to the *cube* (third power) of the wind speed. If the wind speed doubles, then the theoretical wind power increases by a factor of 8. If the wind speed triples, then the theoretical wind power becomes 27 times as great! Note also that the available wind power is proportional to the *square* of the blade length, if all other factors are held constant. That's because the area of a circle (the shape of the region swept out by the blades as they rotate) is proportional to the square of its radius.

Quiz

This is an "open book" quiz. You may refer to the text in this chapter. A good score is eight correct. Answers are in the back of the book.

1. A stand-alone small-scale hydropower system works best
    a. with a power inverter and battery.
    b. with a large dam and a large reservoir.
    c. when used with multiple reservoirs.
    d. where littoral currents run perpendicular to the shore.

2. A tidal turbine can be used to drive a generator by taking advantage of
    a. wave action.
    b. littoral currents.
    c. impoundment.
    d. Any of the above

3. Which of the following does *not*, in theory, affect the amount of electrical power (in kilowatts or megawatts) produced by a large wind turbine when it is properly operating at any given moment in time?
    a. The blade length
    b. The wind speed
    c. The capacity factor
    d. The size of the generator

4. Suppose you are told that a certain wind turbine operates at peak efficiency when the wind speed is 15 m/s. This is equivalent to about
    a. 6.7 mi/h.
    b. 7.7 mi/h.
    c. 29 mi/h.
    d. 34 mi/h.

5. Which of the following is an ocean-derived, exploitable source of electric energy?

    a. The swells produced by distant storms

    b. The undertow produced by breakers

    c. Rip currents

    d. All of the above

6. An impoundment hydroelectric facility is most feasible where

    a. the terrain is nearly level, and a shallow river flows through it.

    b. the terrain is irregular, and a river flows through it.

    c. a river empties into the sea in a large delta.

    d. ocean swells are consistent and large, and rip currents are strong.

7. Suppose a small wind generator produces a constant electrical output of 600 W in a wind that blows at 10 m/s. How much electrical energy does this generator produce in 1 min in a wind that blows at 10 m/s?

    a. 10 J

    b. 60 J

    c. 600 J

    d. 36,000 J

8. Which of the following statements, a, b, or c, if any, is false?

    a. In an interactive wind-power system with batteries, battery power is sold to the electric utilities if the wind turbine provides insufficient power to charge them.

    b. A large-scale wind turbine typically rotates at constant speed for all winds within the operating range.

    c. Furling can help protect small-scale wind turbines from damage in high winds.

    d. All of the above statements are true.

9. An air turbine would likely be found as a component of

    a. a tidal barrage.

    b. a diversion hydroelectric system.

    c. an impoundment hydroelectric system.

    d. a wave-electric generator.

10. A large impoundment dam can cause

    a. excessive oxygen to be dissolved in the water behind the dam.

    b. shortages of water upstream that would not occur if the dam did not exist.

    c. problems for wildlife in the river and reservoir near the dam.

    d. All of the above

CHAPTER 12

Electricity from Atoms and the Sun

Radiant energy can come from within the earth as well as from outer space. For the production of electricity, the most common terrestrial source of radiant energy is *uranium fission*, and the most common extraterrestrial source is the visible light (and to some extent IR and UV) that comes from the sun. *Hydrogen fusion*, while promising, is still in the research-and-development phase.

Atoms

All matter is made up of countless tiny, dense particles. Matter is mostly empty space, but it seems "continuous" because the particles are submicroscopic and move incredibly fast. Each chemical *element* has its own unique type of particle, known as its *atom*. The slightest change in the internal structure of an atom can make a tremendous difference in its outward behavior.

PROTONS, NEUTRONS, AND ATOMIC NUMBERS

The part of an atom that gives an element its identity is the *nucleus*, which is made up of two types of particles, the *proton* and the *neutron*. A teaspoonful of either of these particles, packed tightly together, would mass thousands of kilograms. Protons and neutrons have just about the same mass, but the proton has a positive electric charge while the neutron has no electric charge.

The number of protons in an element's nucleus, the *atomic number*, gives that element its identity. The element with one proton in its nucleus is *hydrogen*; the element with two protons in the nucleus is *helium*. If there are three protons, we have *lithium*, a light metal that combines easily with many other elements. The element with four protons is *beryllium*, also a metal. In general, as the number of protons in an element's nucleus increases, the number of neutrons also increases, although there are some exceptions. Elements with high atomic numbers such as *uranium* (atomic number 92) are therefore more dense than elements with low atomic numbers such as *carbon* (atomic number 6).

ISOTOPES

For any particular chemical element, the number of protons is always the same, but the number of neutrons can vary. Differing numbers of neutrons result in different *isotopes* for a given element. All elements have at least two isotopes, and some have dozens.

Every element has one particular isotope that is most often found in nature. A change in the number of neutrons in the nucleus of an atom results in a change in the mass of that atom. Some isotopes of certain elements stay the same over long periods of time; these are called *stable* isotopes. Others isotopes are not so well-behaved. Their nuclei tend to fall apart, or *decay*. When atomic nuclei decay spontaneously, they are said to be *unstable*. The decay process is always attended by the emission of energy and/or subatomic particles. These emissions are known as *radioactivity*. All the known isotopes of uranium are unstable. Some decay more rapidly than others.

ATOMIC WEIGHT

The *atomic weight* of an element is approximately equal to the sum of the number of protons and the number of neutrons in the nucleus. The most common naturally occurring isotope of carbon has an atomic weight of 12, and is called carbon-12 or C-12. The nucleus of a C-12 atom has six protons and six neutrons. Some atoms of carbon have eight neutrons in the nucleus rather than six. This type of carbon atom has an atomic weight of 14, and is known as carbon-14 or C-14. Uranium can exist in the form of U-234 (atomic weight 234) or U-235 (atomic weight 235) as well as the most common isotope, U-238 (atomic weight 238). In a nuclear power plant, U-235 is the isotope of choice because it has properties that make it possible to induce controlled decay, producing a steady supply of usable energy.

IONS

If an atom has more or less electrons than protons, that atom acquires an electrical charge. A shortage of electrons results in a positive charge; an excess of electrons gives a negative charge. The element's identity remains the same, no matter how great the excess or shortage of electrons. A charged atom is called an *ion*. When a substance contains many ions, the material is said to be *ionized*. Ionized materials generally conduct electricity well, even if the substance is normally not a good conductor.

An element can be both an ion and an isotope different from the usual isotope. For example, an atom of carbon might have eight neutrons rather than the usual six, thus being the isotope C-14, and it might have been stripped of an electron, giving it a positive unit electric charge and making it an ion.

Power from Uranium Fission

The term *fission* means "splitting apart." In *nuclear fission*, the splitting apart of atomic nuclei produces smaller nuclei and different elements. This can happen with many elements. In a nuclear power plant, nuclei of U-235 are split apart deliberately in a regulated fashion. This is called *induced fission*.

THE URANIUM FISSION PROCESS

The key to U-235 fission is bombardment by high-speed neutrons. When a neutron strikes a U-235 atomic nucleus, that nucleus splits almost instantly into two lighter nuclei. As this happens, two or three neutrons are emitted along with *gamma rays*, which are similar to *X rays* but even more penetrating and energetic. Heat energy is also produced, warming up the uranium. If one of the emitted neutrons hits another U-235 nucleus, then that nucleus splits, and the process is repeated. In a large enough sample of U-235, the result of all this nucleus splitting and internal neutron bombardment is a *chain reaction*. Eventually, if the chain reaction goes on long enough, all the U-235 ends up getting split down into nuclei of lighter elements. Then we have *spent nuclear fuel*.

One of three states can prevail when U-235 is subjected to neutron bombardment:

- **Subcritical state**   The reaction dies down before much fuel is spent. This happens if, on the average, less than one emitted neutron strikes another U-235 nucleus and splits it.

- **Critical state**   The reaction sustains itself in a steady fashion until the fuel is spent. This happens if, on the average, one emitted neutron strikes another U-235 nucleus and splits it.

- **Supercritical state**   The reaction increases in intensity, and the uranium heats up to the point that it melts. This happens if, on the average, more than one emitted neutron strikes another U-235 nucleus and splits it.

HOW A FISSION POWER PLANT WORKS

In order for a nuclear fission reactor to work properly, it must be maintained in the critical state. This involves controlling the temperature of the U-235, as well as starting out with a sample of U-235 having exactly the right mass and shape. To some extent, the amount of neutron radiation within the sample can be controlled to keep the system operating in a steady state. When controlled properly, a uranium fission reaction can provide large quantities of usable heat energy for long periods of time.

There is no risk of explosion with U-235 that has been refined specifically for use in nuclear reactors. If you've seen a movie in which a fission reactor blew up like an atomic bomb, it was not based on reality. But there's plenty to worry about if a fission reaction gets out of control. If a reactor is allowed to "go supercritical," the uranium will melt because of the excessive heat. That's a condition called *meltdown*. Reactor-grade U-235 is not refined enough to undergo the sort of split-second, violent chain reaction that occurs in a nuclear bomb, but meltdown can contaminate the soil, water, and air with radioactive isotopes.

A uranium fission reactor is housed in a multilayered structure to keep radiation from escaping and to physically protect it from damage that could arise from external causes. A *radiation shield* and *containment vessel* prevent the escape of radiation or radioactive materials into the surrounding environment. The entire assembly is housed in a massive reinforced building called the *secondary containment structure*. This building, characteristically shaped like a dome or half-sphere for maximum structural integrity, is designed to withstand catastrophes such as tornadoes, hurricanes, earthquakes, and direct hits by aircraft or missiles.

Figure 12-1 is a simplified functional diagram of a nuclear fission power plant. Heat from the reactor is transferred to a water boiler by means of heat-transfer fluid (coolant). The coolant passes from the shell of the boiler back to the reactor through the coolant pump. The water in the boiler is converted to steam, which drives a turbine. The turbine rotates the shaft of an electric generator that is connected into

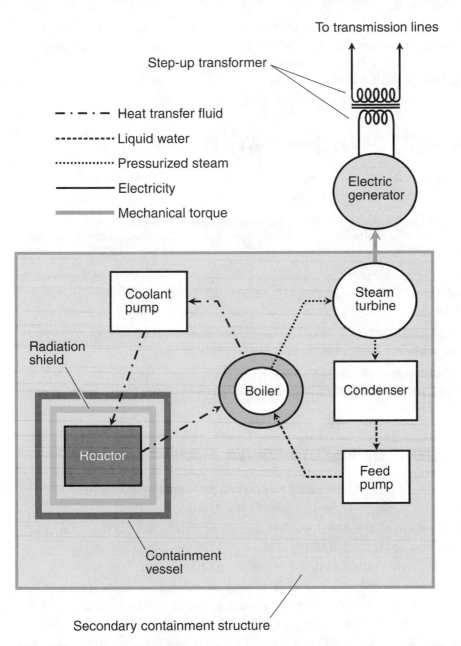

To transmission lines

Step-up transformer

Heat transfer fluid
Liquid water
Pressurized steam
Electricity
Mechanical torque

Electric generator

Coolant pump

Radiation shield

Boiler

Reactor

Steam turbine

Condenser

Feed pump

Containment vessel

Secondary containment structure

Figure 12-1 Simplified functional diagram of a nuclear fission power-generating system, showing one reactor and one turbine.

the utility grid through a step-up transformer. After passing through the turbine, the steam is condensed and sent back to the boiler by the feed pump. The water and heat-transfer fluid are entirely separate, closed systems. Neither comes into direct physical contact with the other. This prevents the accidental discharge of radiation into the environment through the water/steam system.

ADVANTAGES OF FISSION POWER PLANTS

- Uranium is a relatively inexpensive fuel. It can be found in regions widely scattered throughout the world.

- Maintenance of fission power plants, while critical, need not be done as frequently as refueling and maintenance operations in conventional power plants.

- Fission reactors and their associated peripherals can operate in the absence of oxygen. That means they can be completely sealed and, if necessary, placed under the ground or under water without ventilating systems.

- Fission power plants do not produce greenhouse-gas emissions, CO gas, or particulate pollutants as do fossil-fuel power plants.

- Fission power plants, if responsibly built and used, can help the world economy wean itself off of its heavy reliance on fossil fuels for the generation of electricity.

LIMITATIONS OF FISSION POWER PLANTS

- Uranium mining and refining can expose personnel to radioactive dust, and can also release this dust into the air and water.

- Fission reactors produce waste that remains radioactive for many years. Existing and proposed disposal processes for this waste are fraught with technical, environmental, and political problems.

- Although the risk of accident or sabotage involving a nuclear reactor is small, the potential consequences—leakage of radioactive material into the environment—are great, and cannot be discounted.

- Transporting fissionable materials to power plants for use, and transporting waste products to nuclear dumps, can never be a perfectly secure business. The consequences of a security breach could be enormous.

- If certain fissionable nuclear waste products get into the wrong hands, nuclear terrorism or blackmail could result.

- The widespread use of fission reactors faces opposition from certain groups because of the above-mentioned negative factors. This has given rise to public apprehension, particularly in the United States, concerning nuclear energy in general.

PROBLEM 12-1
What is the difference between reactor-grade uranium and weapons-grade uranium?

SOLUTION 12-1
For use in a fission reactor, uranium must be refined, or *enriched*, until it contains at least 3 percent U-235. In order to make fuel for an atom bomb, uranium must be enriched until it contains at least 90 percent U-235.

Power from Hydrogen Fusion

Because of public concerns about nuclear fission in the United States, hydrogen fusion has been suggested as a way to take advantage of the properties of the atom in order to generate electricity. In theory, this is a great idea. Hydrogen fusion is more efficient in converting matter to energy than fission, and no radioactive waste is produced. But a workable hydrogen fusion reactor has not yet been developed.

FUSION IN THE SUN

Physicists believe that the sun converts hydrogen to helium by means of nuclear fusion. The term "fusion" means "combining." Hydrogen fusion requires extremely high temperature. The powerful gravitation imposed by the sun's huge mass keeps the core in a constantly compressed state. This compression keeps the core hot enough for hydrogen fusion to occur.

Solar hydrogen fusion is a multistep process. At first, two hydrogen nuclei (protons) are squeezed together, emitting a *positron*, also known as an *anti-electron*. A positron has the same mass as an electron, but carries a unit positive charge rather than a unit negative charge. A *neutrino* is also emitted. Neutrinos are something like electrons with no electric charge and the ability to penetrate matter to an incredible extent. The fusing of two protons is attended by a loss of a unit positive charge; as a result, one of the protons becomes a neutron. This produces a nucleus of *deuterium* (H-2), a heavy isotope of hydrogen consisting of one proton and one neutron. The deuterium nucleus combines with another proton to form a nucleus of *helium-3*

(He-3), containing two protons and one neutron. As this happens, a burst of gamma radiation is emitted. Two He-3 nuclei, resulting from two separate iterations of the above-described process, then combine to form a nucleus of *helium-4* (He-4), which has two protons and two neutrons. This is the isotope of helium we use to fill up lighter-than-air balloons. In this final phase, two protons are ejected. These can contribute to further fusion reactions.

In the solar fusion process (see Figure 12-2), the total mass of the matter produced is a little less than the total mass of all the ingredients. The "missing mass" is converted into energy according to the famous Einstein equation:

$$E = mc^2$$

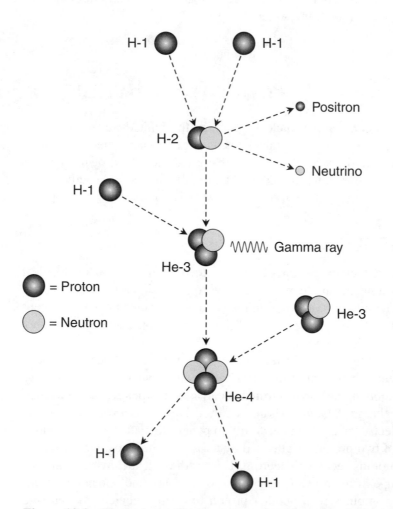

Figure 12-2   The hydrogen fusion process that takes place in the core of the sun.

where E is the energy in joules, m is the "missing mass" in kilograms, and c is the speed of light, equal to approximately 3×10^8 meters per second. The sun produces a tremendous amount of energy in this way, because hydrogen nuclei are converted to helium nuclei continuously and in vast numbers. There is enough matter in the sun to keep its hydrogen fusion process going for millions of millennia yet to come. Eventually the hydrogen fuel supply will run out, but not in your lifetime or mine!

FUSION IN BOMBS

In a *hydrogen bomb*, a different hydrogen fusion reaction takes place. This mode, if it can ever be controlled, may be used in a fusion reactor. Instead of simple hydrogen nuclei, which are protons, nuclei of *heavy hydrogen* merge. One nucleus is of deuterium (H-2), consisting of one proton and one neutron. The other nucleus is of *tritium* (H-3), which contains one proton and two neutrons. When these combine, the result is a nucleus of He-4, with the extra neutron ejected (see Figure 12-3). Along with this, energy is liberated, just as is the case inside the sun. For this mode, called *deuterium-tritium fusion* or *D-T fusion*, deuterium and tritium fuel must be supplied. Ordinary hydrogen (H-1) won't work. Several other fuel combinations can theoretically work for nuclear fusion, but the D-T mode has received the most attention.

In the sun, the fusion process goes on continuously because of the heat produced by the crushing pressure of gravitation, and also because of the heat generated from the reactions themselves. In a hydrogen bomb, the necessary heat to start the reaction is supplied by a fission bomb, but the reaction burns itself out in a hurry. It's

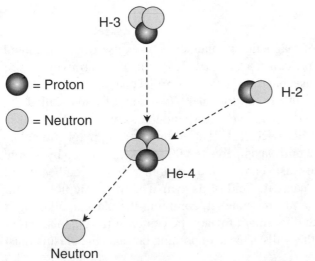

Figure 12-3 The fusion process that occurs in a hydrogen bomb.

impossible to get enough gravitation to start and maintain fusion indefinitely and in a controlled manner in a terrestrial sample of simple hydrogen, deuterium, or tritium. In order to make fusion work for the purpose of generating useful power, the fuel must somehow be confined.

PLASMA FUEL

When a sample of gas is heated to an extremely high temperature, the electrons are stripped away from the nuclei. The atoms therefore become ions. Instead of "orbiting" the positively charged nucleus of a specific atom, an electron is "free" to move from atom to atom, or even to travel through space all by itself. When this ionization happens, the gas, which is normally a poor conductor of electric current, becomes a good conductor. A substance in this state is known as a *plasma*. Because a plasma differs from an ordinary gas, the plasma state has been called the *fourth phase of matter* (the other three being solid, liquid, and gas). In the prototypes of fusion reactors that most scientists favor, deuterium and tritium exist in the plasma state, heated to temperatures comparable to those in the cores of stars.

The behavior of a plasma can be dramatically affected by external electric or magnetic fields. An electric or magnetic field can cause a plasma to constrict, distort, bunch up, or spread out. If a plasma is surrounded by an external electric or magnetic field having certain properties, the plasma can be kept within a small, defined space, even if it becomes hot enough to sustain hydrogen fusion reactions. The external fields act on the plasma in much the same way as gravitation inside the sun keeps the hot core gases confined. The use of external magnetic fields to compress and hold a hot D-T plasma in place during a fusion reaction is known as *magnetic confinement*.

THE TOKAMAK

The most promising method of magnetic confinement makes use of an evacuated toroidal (donut-shaped) enclosure called a *tokamak*. This term is an acronym derived from a Russian descriptive phrase that translates as "toroidal chamber and magnetic coil." The plasma is contained inside the tokamak. Two sets of coils, called the *toroidal field coils* and the *poloidal field coils*, surround the toroidal enclosure (see Figure 12-4). The coils carry electric currents that produce strong magnetic fields. An electric current of up to five million amperes (5,000,000 A), provided by a large transformer, travels through the plasma around the toroid in a circular, endless loop. This plasma current creates a magnetic field of its own. The magnetic fields from the currents in the coils and the plasma interact, confining the plasma, aligning it within the tokamak chamber, and forcing it toward the center of the chamber cross section, keeping it away from the walls. This is important, because the plasma must

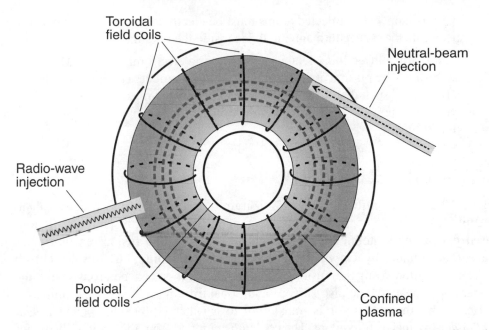

Toroidal
field coils

Neutral-beam
injection

Radio-wave
injection

Poloidal
field coils

Confined
plasma

Figure 12-4   Functional diagram of a tokamak, showing plasma confinement coils and two methods of heating the plasma.

be heated to more than 100 million degrees Celsius (100,000,000°C) in order for fusion to occur! If the superheated plasma were to contact the tokamak wall at any point, the chamber would rupture, air would leak in, the plasma would cool below the critical temperature, and the fusion reaction would cease. The helium product of the fusion process is removed from the chamber by *divertors*.

HEATING THE PLASMA

Several processes can be implemented in order to obtain the high plasma temperature necessary to sustain the fusion reaction:

- **Ohmic heating**   Arises from the fact that the current, as it circulates in the plasma, encounters a finite resistance. This means that a certain amount of power is dissipated in the plasma, just as a wire gets hot when it carries high current.

- **Self heating**   Takes place because the fusion reaction produces heat itself, and some of this heat is absorbed by the plasma.

- **Neutral-beam injection**   Involves firing high-energy beams of neutral H-2 and H-3 atoms into the plasma. These heat the plasma when they collide with its atoms. The injected atoms must be electrically neutral so they can penetrate the powerful magnetic fields inside the tokamak.

- **Radio-wave injection**   Done by transmitting electromagnetic (EM) waves into the plasma at several points in the chamber. These waves have a frequency such that their energy is absorbed by the plasma. Figure 12-4 shows one point of neutral-beam injection and one point of radio-wave injection.

GETTING THE ELECTRICITY

Figure 12-5 is a simplified functional diagram of a hydrogen fusion power plant. Except for the nature of the reaction, this type of power plant resembles a fission-based generating system. The plasma chamber, where the fusion reaction takes place, is surrounded by a *moderator*, which consists of lithium blankets that absorb neutron radiation from the fusion reaction. The high-speed neutrons cause the moderator to heat up, and also "breed" additional tritium fuel from the lithium.

Heat from the moderator is transferred to a water boiler, also called a *heat exchanger*, by means of coolant. The coolant is pumped from the shell of the boiler back to the plasma chamber. The water in the boiler is converted to steam, which drives a turbine. After passing through the turbine, the steam is condensed and sent back to the boiler by a feed pump. The turbine rotates the shaft of an electric generator that is connected into the utility grid through a step-up transformer.

ADVANTAGES OF HYDROGEN FUSION FOR POWER

- The only material byproducts of hydrogen fusion are He-4, a harmless gas, and tritium that can serve as additional fuel.

- Deuterium fuel can be easily obtained from water. Lithium is abundant in the earth's crust. Tritium can be bred in the reactor. These are the only material ingredients necessary to operate a D-T fusion reactor.

- A hydrogen fusion power plant will produce no greenhouse-gas emissions, CO gas, or particulate pollutants as do fossil-fuel power plants.

- A working fusion reactor will be safer than a fission reactor. Meltdown will not occur if the reactor is damaged, because terrestrial fusion reactions cannot be sustained without the continuous infusion of fuel and energy.

- Terrestrial fusion is not a chain reaction. It can't get out of control and cause a fusion reactor to blow up. The reason a hydrogen bomb explodes is because abundant fuel is supplied and it is used up almost instantaneously,

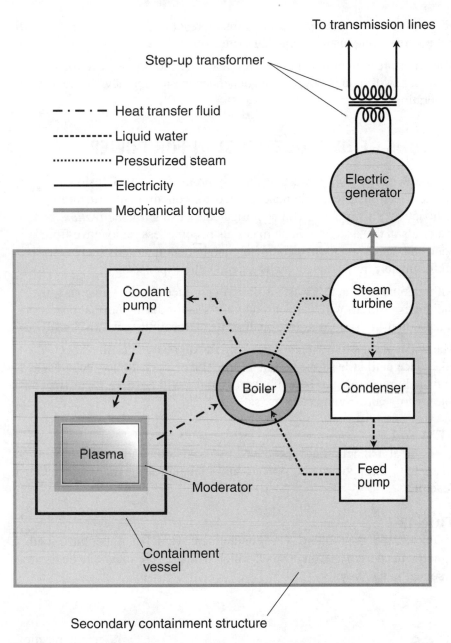

Figure 12-5 Simplified functional diagram of a hydrogen fusion power-generating system, showing one reactor and one turbine.

not because of a chain reaction. In a fusion reactor, the quantity of fuel will not be great enough to generate an explosion.

- The widespread deployment of fusion reactors, should they ever be perfected, will reduce or eliminate dependence on nonrenewable fuels for generating electricity.

LIMITATIONS OF HYDROGEN FUSION FOR POWER

- Although no radioactive waste is directly produced by D-T fusion, the emitted neutrons eventually make the reactor containment structure radioactive. This problem can be mitigated by using *low-activation* materials in the structure. Such materials become less radioactive from neutron bombardment than common containment materials such as steel. Unfortunately, low-activation alloys tend to be expensive.

- Although no radioactive waste is directly produced by D-T fusion, some radioactive tritium will be released by the reactor during normal operation. It has a half-life (that is, it loses half of its radioactivity) in 12 years.

- The widespread deployment of working fusion reactors is not expected to take place until at least the middle of the 21st century. Major technical and logistic hurdles remain. In addition, the public will have to be convinced that hydrogen fusion reactors are safe.

PROBLEM 12-2
Will the powerful magnetic fields produced by the field coils, and by the plasma, pose a danger to the people who operate and maintain a magnetic-confinement fusion reactor?

SOLUTION 12-2
This should not be a problem. The containment vessel can be lined with a ferromagnetic metal such as iron or steel, which will keep the magnetic fields away from personnel in the vicinity.

Photovoltaics

A *photovoltaic* (PV) *cell* is a specialized form of *semiconductor diode* that converts visible light, IR radiation, or UV radiation directly into electricity. When used to obtain electricity from sunlight, this type of device is known as a *solar cell*. One of the most common types of solar cell is made of specially treated silicon, and is known as a *silicon PV cell*.

STRUCTURE AND OPERATION

The basic structure of a silicon PV cell is shown in Figure 12-6. It is made out of two types of silicon, called *P type* and *N type*. The heart of the device is the surface at which these two types of materials come together, known as the *P-N junction*. The top of the assembly is transparent so that light can fall directly on the junction. The positive electrode is made of metal *ribbing* interconnected by tiny wires. The negative electrode is a metal base called the *substrate*, which is placed in contact with the N-type silicon.

When radiant energy strikes the P-N junction, a voltage or *potential difference* develops between the P-type and N-type materials. When a load is connected to the cell, the intensity of the current through the load increases in proportion to the brightness of the light striking the P-N junction of the cell—up to a certain critical point. As the light intensity increases past this critical point, the current levels off at

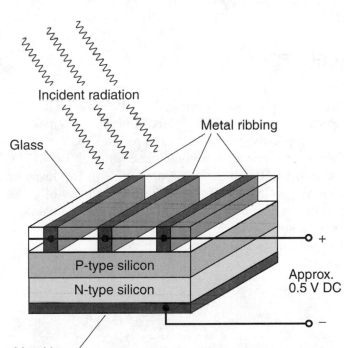

Figure 12-6 Construction of a silicon photovoltaic cell.

a maximum called the *saturation current*. The ratio of the available output power to the light power striking a PV cell is called the *efficiency*, or the *conversion efficiency*, of the cell.

VOLTAGE, CURRENT, AND POWER

Most silicon solar cells provide about 0.5 V DC when there is no load connected. If there is very low current demand, even fairly dim light can generate the full output voltage (symbolized V_{out}) from a PV cell. As the current demand increases, brighter light is needed to produce the full V_{out}.

There is a maximum limit to the current that can be provided from a PV cell, no matter how intense the incident radiation. This is called the *maximum deliverable current* and can be symbolized as I_{max}. The I_{max} value for a PV cell depends on the surface area of the P-N junction and on the technology used in the manufacture of the device.

The maximum deliverable power (P_{max}) for a silicon PV cell, in watts, is equal to the product of V_{out} in volts and I_{max} in amperes. Thus:

$$P_{max} = 0.5\, I_{max}$$

SERIES-PARALLEL CONNECTIONS

Photovoltaic cells are often combined in *series-parallel* to obtain higher power output than is possible with a single cell alone. When multiple PV cells (or parallel combinations of cells) are connected in series, the V_{out} values add up. When multiple PV cells (or series combinations of cells) are connected in parallel, the I_{max} of the whole set is equal to the I_{max} of a single cell or combination, multiplied by the number of cells or combinations. The P_{max} of a series-parallel PV matrix made of identical cells is equal to the P_{max} of each cell, times the total number of cells. Alternatively, P_{max} is equal to the product of V_{out} and I_{max} for the complete matrix.

As an example, consider 10 parallel-connected sets of 36 series-connected PV cells, assuming I_{max} = 2.2 A for each series-connected set. We can calculate the following values for the matrix:

$$V_{out} = 36 \times 0.5\text{ V}$$

$$= 18\text{ V}$$

$$I_{max} = 10 \times 2.2\text{ A}$$

$$= 22\text{ A}$$

$$P_{max} = V_{out} I_{max}$$

$$= 18 \text{ V} \times 22 \text{ A}$$

$$= 396 \text{ W}$$

This can be rounded off to 400 W. It is a theoretical value, however, not a real-life result. When a load is connected to a PV system, V_{out} is slightly lower than the theoretical amount. When numerous PV cells are connected in series, there is a voltage drop of several percent under load as a result of the internal resistance of the combination. In the above case, V_{out} might be only 14 V when the current demand is near I_{max}. Therefore, the actual value of P_{max} would be

$$P_{max} = 14 \text{ V} \times 22 \text{ A}$$

$$= 308 \text{ W}$$

This can be rounded off to 300 W.

LOW, MEDIUM, AND HIGH VOLTAGE

In low-voltage, low-current PV systems, individual cells are normally connected in series to obtain the desired V_{out}. For charging a 12-V battery, a common V_{out} value is 16 V, requiring 32 cells in series. Such a series-connected set is called a *PV module*. In order to get higher I_{max}, multiple modules can be connected in parallel to form a *PV panel*. Finally, if even higher V_{out} or I_{max} is necessary, multiple panels can be connected in series or parallel to obtain a *PV array*.

Although high voltages (such as 5 kV DC) can theoretically be obtained by connecting hundreds or even thousands of PV cells in series, this is problematic because the *internal resistances* of the cells add up, severely limiting I_{max} and causing the output voltage to drop under load. This can be overcome in high-power PV applications by connecting a large number of cells or low-voltage modules in parallel, making many identical such sets, and then connecting all the parallel sets in series.

If medium voltage is needed from a low-voltage solar panel, a power inverter can be used along with a high-capacity rechargeable battery. The solar panel keeps the battery charged; the battery delivers high current on demand to the power inverter. Such a system provides common 117-V AC electricity from a 12-V DC or 24-V DC source.

MOUNTING AND LOCATION

A solar panel produces the most power when it is broadside to incident sunlight, so the sun's rays shine straight down on the surface. However, this orientation is not critical. Even at a slant of 45 degrees (45°) with respect to the sun's rays, a solar panel receives 71 percent as much energy per unit surface area as it does when optimally aligned. A misalignment of up to 15° makes almost no difference.

Solar panels should be located where they will receive as much sunlight as possible, averaged out during the course of the day and the course of the year. Mountings should be sturdy enough so the panels will not be ripped loose or forced out of alignment in strong winds. One of the most popular arrangements involves mounting a solar panel, or a set of panels, directly on a steeply pitched roof that faces toward the south.

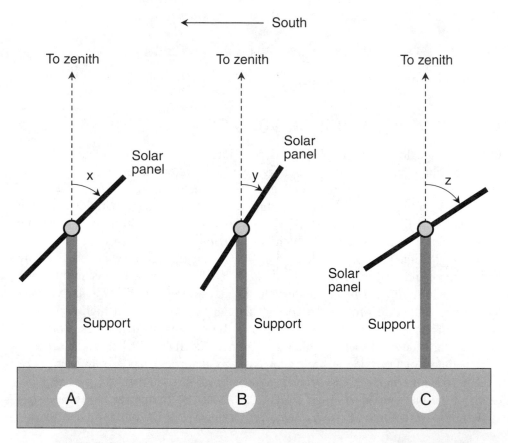

Figure 12-7 Optimal placement of fixed, south-facing solar arrays for locations in northern temperate latitudes for year-round operation (A), low-solar-angle-season operation (B), and high-solar-angle-season operation (C). The variables x, y, and z represent angles in degrees with respect to the zenith. In each case the panel is viewed edge-on, looking west.

The ideal bearing arrangement for a solar panel would be a motor-driven *equatorial mount*, similar to those used with sophisticated telescopes. This would allow the panel to follow the sun all day, every day of the year. However, such a system is impractical for most people, and the cost would be prohibitive for large panels or multipanel arrays. The next best thing is a mount with a single bearing that allows for the panel to be manually tilted, always facing generally south in the Northern Hemisphere or generally north in the Southern Hemisphere. This type of system is shown in Figures 12-7 and 12-8 for a single panel. Angles x, y, and z represent the tilt of the panel surface with respect to the zenith.

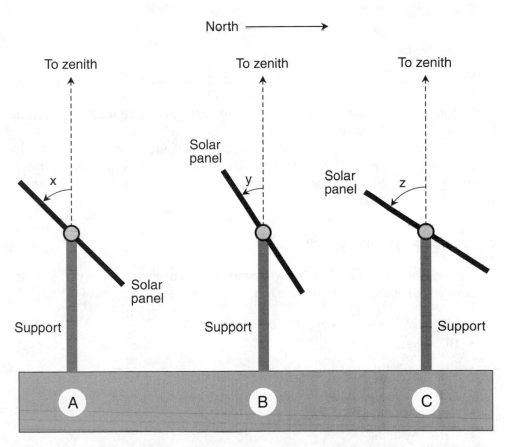

Figure 12-8 Optimal placement of fixed, north-facing solar arrays for locations in southern temperate latitudes for year-round operation (A), low-solar-angle-season operation (B), and high-solar-angle-season operation (C). The variables x, y, and z represent angles in degrees with respect to the zenith. In each case the panel is viewed edge-on, looking west.

NORTHERN HEMISPHERE

The arrangements in Figure 12-7 will work between approximately 20° north latitude (20° N) and 60° north latitude (60° N). Cities in such locations include

- Las Vegas, USA
- Chicago, USA
- Miami, USA
- Paris, France
- Berlin, Germany
- Moscow, Russia
- Beijing, China
- Osaka, Japan

In Figure 12-7A, a year-round panel position is shown. The angle x should be set to 90° minus the north latitude at which the system is located.

If an adjustable bearing is provided, two tilt settings can be used as shown in Figure 12-7B and C. From September 21 through March 20 (autumn and winter), the arrangement shown at B is optimal, and the angle y should be set to approximately 78° minus the latitude. From March 21 through September 20 (spring and summer), the arrangement shown at C is optimal, and the angle z should be set to approximately 102° minus the latitude.

SOUTHERN HEMISPHERE

The arrangements in Figure 12-8 will work between approximately 20° south latitude (20° S) and 60° south latitude (60° S). Cities in such locations include

- Santiago, Chile
- Buenos Aires, Argentina
- Rio de Janeiro, Brazil
- Cape Town, South Africa
- Durban, South Africa
- Perth, Australia
- Sydney, Australia
- Auckland, New Zealand

In Figure 12-8A, a year-round panel position is shown. The angle x should be set to 90° minus the south latitude at which the system is located.

If an adjustable bearing is provided, two tilt settings can be used as shown in Figure 12-8B and C. From March 21 through September 20 (autumn and winter), the arrangement shown at B is optimal, and the angle y should be set to approximately 78° minus the latitude. From September 21 through March 20 (spring and summer), the arrangement shown at C is optimal, and the angle z should be set to approximately 102° minus the latitude.

PROBLEM 12-3

Suppose that a large number of solar modules are manufactured, each consisting of 53 identical PV cells in series. Each PV cell provides exactly 0.5 V at exactly 2 A in bright sunlight. You make two panels, each consisting of 20 of these modules in parallel. You connect these two panels in series. Suppose that the manufacturer of the modules tells you that under load, the output voltage can be expected to drop by 10 percent compared with the theoretical (no-load) value. What are the practical values of V_{out}, I_{max}, and P_{max} for this array?

SOLUTION 12-3

Let's calculate the *theoretical output voltage* (call it V_{out-th}) and then reduce this figure by 10 percent to get the actual output voltage, V_{out}. Then we'll calculate I_{max} for the whole array. Finally, we'll determine P_{max} by finding the product of V_{out} and I_{max}.

Each module theoretically provides 53×0.5 V $= 26.5$ V. The parallel combination of 20 modules to form a panel also produces a theoretical output voltage of 26.5 V. (Identical voltages in parallel do not add up.) Two such panels are connected in series to form the array. Thus:

$$V_{out-th} = 2 \times 26.5 \text{ V}$$

$$= 53 \text{ V}$$

Reducing this by 10 percent is the equivalent of multiplying it by 90 percent, or 0.9. Therefore:

$$V_{out} = 0.9 \times 53 \text{ V}$$

$$= 47.7 \text{ V}$$

Each cell provides 2 A of maximum deliverable current. Therefore, each series-connected set, forming a module, also provides 2 A of maximum deliverable current. (Identical currents in series do not add up.) When 20 modules are connected in parallel to form a panel, the maximum deliverable current is 20×2 A $= 40$ A.

When two of these panels are connected in series to form the final array, the resulting maximum deliverable current is still 40 A. Therefore, for the entire array:

$$I_{max} = 40\ A$$

In order to find the maximum deliverable power, we multiply V_{out} by I_{max}, as follows:

$$P_{max} = V_{out}\, I_{max}$$
$$= 47.7\ V \times 40\ A$$
$$= 1908\ W$$

This can be rounded off to 1900 W or 1.9 kW.

Large-Scale PV Systems

A large-scale PV system, designed to provide power to many users, is sometimes called a *solar farm*. The heart of this type of power plant is a massive array of PV cells. Solar farms can be found scattered around the sunniest parts of the United States and several other nations. Depending on the size of the array, a solar farm can furnish anywhere from a few dozen kilowatts up to a hundred megawatts (100 MW) or more.

HOW IT WORKS

A large solar farm typically has thousands (and in some cases hundreds of thousands) of individual PV cells connected in a complex web of modules, panels, and arrays. In the most sophisticated large-scale solar farms, the arrays are set on equatorial mounts so they can be turned directly toward the sun during all the hours of daylight. A computer-controlled mechanical system guides the bearings so the adjustment is optimum for every day of the year. The PV arrays are connected to the utility grid through power inverters and transformers. Figure 12-9 shows the basic configuration. In the biggest solar farms, many inverters are connected in tandem, and their waves kept in phase. The combination feeds a step-up transformer that runs to a high-voltage AC transmission line.

ADVANTAGES OF LARGE-SCALE PV SYSTEMS

- Sunlight is a renewable resource, and the supply is practically unlimited.
- Solar farms generate no greenhouse gases, toxic compounds, or particulate matter.

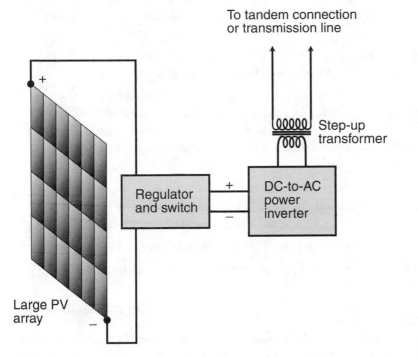

Figure 12-9   Simplified functional diagram of a solar-electric generator using a large PV array, an inverter, and a step-up transformer for connection into the utility grid.

- In operation, PV cells make no noise.
- Solar farms, like wind farms, can be dispersed over wide regions. This distributes the energy source, an attribute that can help in the evolution of a fault-tolerant utility grid.
- Solar power can be used to supplement other modes of electric power generation. This increases the diversity of a nation's electrical system.
- A typical solar farm has a low profile. There are no large towers or buildings. A solar farm is as unobtrusive as a grove of fruit trees.
- Solar farms are an important part of the long-term quest to reduce or eliminate dependence on nonrenewable fuels for generating electricity.

LIMITATIONS OF LARGE-SCALE PV SYSTEMS

- Large-scale PV systems are not cost-effective in places that get relatively little direct sunlight.

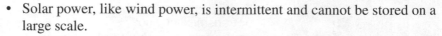

- Solar power, like wind power, is intermittent and cannot be stored on a large scale.

- Solar power cannot, by itself, totally satisfy the electrical needs of a city, state, or nation. It is at best a supplemental power source.

- Unless proper electronic design precautions are taken and the site is chosen carefully, *load imbalance* can occur if part of a solar array is in sunlight while another part is in shadow. This reduces the output of the array more than would be expected by calculating the percentage of the surface area that is shaded.

- In order to get optimum performance from a large-scale PV system, the panels must be mounted on movable bearings. This can be costly. A fixed arrangement, no matter how well thought-out, is a compromise.

- Solar farms require a certain amount of dedicated real estate.

- Hail and wind storms can damage or destroy solar modules, panels, and arrays.

PROBLEM 12-4

What, exactly, is silicon, from which most PV cells are made? Is the supply of this material limited? Should we worry about running out someday?

SOLUTION 12-4

Silicon is an element with atomic number 14 and atomic weight 28. In its pure state, silicon is a lightweight, rather brittle metal, similar in appearance to aluminum. Silicon is a *semiconductor* substance; it conducts electric currents better than an electrical insulator, but not as well as excellent conductors such as silver and copper. Silicon is found in great quantities in the crust of the earth. There's enough of it to supply the needs of humanity indefinitely. In its natural state, silicon is almost always combined with oxygen or other elements. A good source is *silicon dioxide*, which is found in quartzite sand.

Small-Scale PV Systems

The term *small-scale* applies to PV systems that can produce enough electricity under ideal conditions to power a residential house. Like their larger counterparts, small-scale PV systems generate power on an intermittent basis. In order to obtain a continuous supply of electricity with a small-scale PV power plant, it is necessary to use storage batteries or an interconnection to the electric utility, or both.

STAND-ALONE SYSTEM

A *stand-alone small-scale PV system* employs rechargeable batteries to store the electric energy supplied by a PV panel or array. The batteries provide power to an inverter that produces 117 V AC. In some systems the battery power can be used directly, but this necessitates the use of home appliances designed for low-voltage DC. Figure 12-10 is a functional block diagram of a stand-alone small-scale PV system that can provide 117 V AC for the operation of small household appliances.

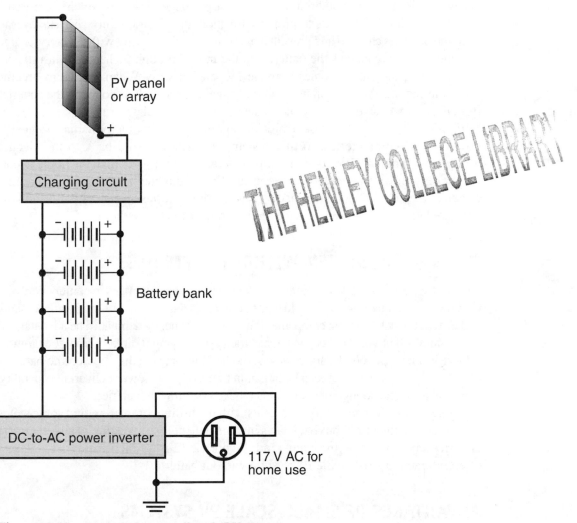

Figure 12-10   A stand-alone small-scale PV system.

The use of batteries allows the system to produce usable power even if there is not enough light for the PV cells to operate. A stand-alone PV system of this type offers independence from the utility companies. However, a blackout will occur if the system goes down for so long that the batteries discharge and there is no backup power source.

INTERACTIVE SYSTEM WITH BATTERIES

An *interactive small-scale PV system with batteries* is similar to a stand-alone system, but with one significant addition. If there is a prolonged spell in which there is not enough light for the PV cells to properly function, the electric utility can take over to keep the batteries charged and prevent a blackout. A switch, along with a battery-charge detection circuit, connects the batteries to the utility through a charger if insufficient power, or no power at all, comes from the PV panel or array. When conditions become favorable and the PV cells can work properly again, the switch disconnects the batteries from the utility charger and reconnects them to the PV panel or array.

Figure 12-11 is a functional block diagram of an interactive small-scale PV power plant with batteries. In this system, power is not sold back to the electric utility, even when the PV panel or array generates more power than the home needs. Power only flows one way, from the utility line to the batteries through a charging circuit and switch. That takes place only when the batteries require charging and the PV panel or array does not provide enough power to charge them.

INTERACTIVE SYSTEM WITHOUT BATTERIES

An *interactive small-scale PV system without batteries* also operates in conjunction with the utility companies. Energy is sold to the companies during times of minimum demand, and is bought back from the companies during times of heavy demand. The advantage of this system is that you can keep using electricity (by buying it directly from the utilities) if there is a long period of dark weather. Another advantage is that, because no batteries are used, this type of system can be larger, in terms of peak power-delivering capability, than a stand-alone arrangement or an interactive system with batteries.

Some states offer good buyback deals with the utility companies, and some states do not. Check the utility buyback laws in your state before investing in an interactive electric-generating system of any kind. Figure 12-12 is a functional block diagram of an interactive small-scale PV system without batteries.

ADVANTAGES OF SMALL-SCALE PV SYSTEMS

- Sunlight is a renewable resource, and the supply is practically unlimited.

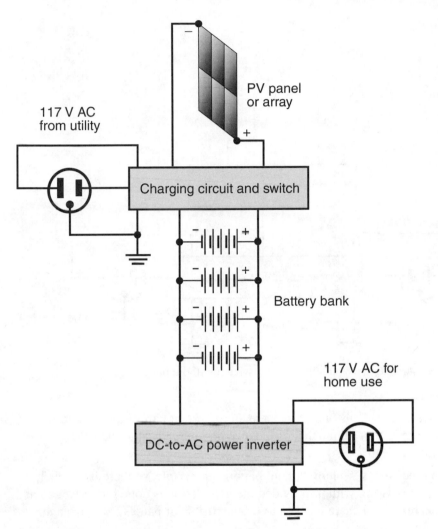

Figure 12-11   An interactive small-scale PV system with batteries.

- Photovoltaic systems generate no greenhouse gases, toxic compounds, particulate matter, or other pollutants.

- In operation, PV cells make no noise.

- Small-scale PV systems can supplement other electric energy sources. The use of a small-scale PV system in conjunction with wind and/or small-scale hydropower makes it easier for a home to operate off the grid than reliance on a single alternative source.

- A typical small-scale PV system has a low profile. In some cases it's hard to tell it's there at all.

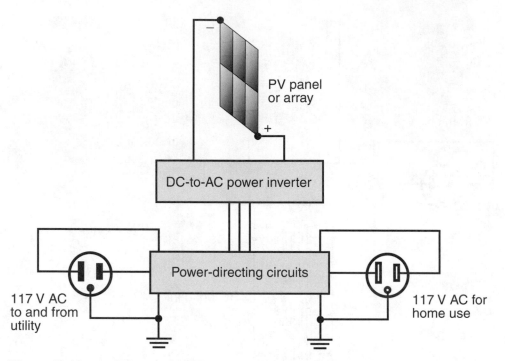

Figure 12-12   An interactive small-scale PV system without batteries.

- Photovoltaic systems of all kinds are an important part of the long-term quest to reduce or eliminate dependence on nonrenewable fuels for generating electricity.

- An interactive PV system without batteries is simple, so there isn't much that can go wrong with it, provided the installation is done properly, and as long as reasonable care is taken to protect the solar panels from damage.

- There is practically no maintenance involved with a PV system that does not use batteries. Once it is up and running, it can be pretty much left alone.

LIMITATIONS OF SMALL-SCALE PV SYSTEMS

- Photovoltaics only provide power to a system when there is enough illumination. Small-scale PV systems rarely justify the cost in places that get relatively little direct sunlight.

- If the solar panels get covered with snow or debris, those materials must be removed manually in order for the system to function.

- Problems with *load imbalance* can occur if part of a solar array is in bright sunlight while another part is in shadow, just as is the case in a large-scale PV system.

- A hail or wind storm can damage or destroy a set of solar panels.

- Care must be taken to ensure that the current demanded from the system does not exceed the maximum deliverable current.

- A small-scale PV system that can supply all the needs of a home, without any sacrifice at all, involves a large up-front cost that may never be recovered.

- In a PV system that uses lead-acid storage batteries, the batteries can produce dangerous fumes. They also require maintenance.

PROBLEM 12-5
In order to keep a stand-alone residential PV system running for prolonged cloudy periods, can't a huge storage battery bank be used—one so large that it can provide the necessary power for days or weeks without any charge from the PV panel or array?

SOLUTION 12-5
In order for this scheme to work, some means would be necessary to ensure that the battery bank would attain a full charge from the PV cells after one or two sunny days. The only way to do that would be to have a huge PV array to match the massive battery. (A huge PV array could also provide limited charging current for the battery bank in gloomy weather.) The oversized battery bank and PV array would translate to an enormous up-front cost. Ventilating and maintaining the battery bank could be problematic. But if you have unlimited financial resources, have access to good engineers, have the required real estate for a big PV array, are determined to get off the electric utility grid, and live in a reasonably sunny location, it is possible to build a medium-scale, stand-alone PV system that will keep your house electrified all the time.

Quiz

This is an "open book" quiz. You may refer to the text in this chapter. A good score is eight correct. Answers are in the back of the book.

1. Suppose a silicon PV cell generates 0.50 V at 2.0 A in direct sunlight. How many watt-hours (Wh) can this cell produce if exposed to direct sunlight for 6 h?

   a. 0.1 Wh

   b. 1.0 Wh

   c. 6.0 Wh

   d. 36 Wh

2. Which of the following substances is radioactive?

   a. H-1

   b. H-2

   c. H-3

   d. He-4

3. A uranium fission reactor cannot explode like an atom bomb because

   a. reactor-grade uranium is not refined enough to undergo the rapid chain reaction that occurs in an atom bomb.

   b. reactor-grade uranium atoms are not of the correct isotope for making bombs.

   c. it is impossible to gather enough uranium together in a fission reactor to cause an atomic explosion.

   d. the temperature in a fission reactor is too high for explosions to occur.

4. If a solar array is in bright sunlight and then suddenly an opaque piece of canvas falls onto it, covering up some of the panels,

   a. the output voltage will suddenly rise.

   b. load imbalance may occur if the array is not properly designed.

   c. the maximum deliverable current of all the cells will rise.

   d. the current output of the entire array will stay the same.

5. A plasma is a good conductor of electric current because

   a. the protons are bound tightly to the nuclei.

   b. the temperature is extremely low.

   c. the substance has low density.

   d. electrons can move easily among atoms.

6. Suppose 34 silicon PV cells are connected in series, and each cell can provide 1.7 A in direct sunlight. Approximately how much current can the module provide in direct sunlight, neglecting the effects of internal resistance?

   a. 0.05 A

   b. 1.7 A

   c. 20 A

   d. 58 A

7. The atomic number for a chemical element is always equal to

   a. the number of neutrons in a single atom.

   b. the number of protons in a single atom.

   c. the sum of the number of protons and the number of neutrons in a single atom.

   d. the number of electrons in a single atom.

8. The most common isotope of uranium has an atomic weight of about

   a. 92.

   b. 234.

   c. 235.

   d. 238.

9. A lithium moderator in a nuclear fusion reactor can be used to

   a. prevent the plasma from becoming ionized.

   b. keep the plasma hot.

   c. breed tritium fuel for the reactor.

   d. no benefit, because lithium is not massive enough.

10. When a uranium fission reaction dies down a long time before all the fuel has been spent, the condition is known as

    a. supercritical.

    b. critical.

    c. subcritical.

    d. quasicritical.

CHAPTER 13

Exotic Electrification Methods

Electricity can be derived from the earth's interior heat, from the combustion or processing of biological waste matter, and from fuel cells. There is also electrical energy stored between the earth and the upper atmosphere.

Geothermal Power

Translated from the Greek, *geothermal* means *earth heat*. The earth's core is incredibly hot, mainly as a result of the decay of radioactive materials with half lives measured in millions or billions (thousand-millions) of years. In fact, if you go deep enough, the rocks in the earth are found in a liquid state called *magma*. Periodically this magma reaches the surface in various places. Then we observe volcanic eruptions.

GO DEEP, GET HEAT

On the average, the temperature of the earth increases by about 28°C for every kilometer, or 80°F for every mile, of depth below the surface for the first several kilometers down. In some locations, the temperature rises faster than this; in some locations it rises more slowly. Wherever you are, if you can sink a well deep enough, you can reach rocks that will boil water and get steam. That's the key to geothermal power generation.

Not surprisingly, the best locations for *geothermal power plants* are places where the temperature rises rapidly with increasing depth. Volcanic regions are excellent. Geologically stable places, or nonvolcanic locations at high elevation, are usually poor sites, but there are exceptions. Certain parts of Wyoming and South Dakota in the United States, for example, are at high elevation, do not have active volcanoes nearby, and yet offer promise for geothermal power generation.

FLASH-STEAM AND DRY-STEAM SYSTEMS

Figure 13-1 is a functional diagram of a *flash-steam geothermal power plant*. Water is forced down into an *injection well* by a *groundwater pump*. The well must be sunk deep enough to reach subterranean rocks at a temperature higher than the boiling point of water. The water filters through the rocks where it becomes heated and rises back up through the nearby *production well*. The hot water from the production well enters a *flash tank* where some of the water boils rapidly into vapor. Water that remains liquid in the flash tank is returned to the groundwater pump to be forced down into the earth again.

The vapor from the flash tank drives a steam turbine, which turns the shaft of an electric generator. After passing through the turbine, the steam is cooled in a *condenser*. This returns the water vapor to the liquid state, and this liquid is forced by the groundwater pump back down into the earth along with the diverted water from the flash tank. Some of the condensed vapor can be used for drinking and irrigation because it is, in effect, distilled. The flash tank must be periodically flushed and cleaned to get rid of mineral buildup. If the water from the production well has high mineral content, the flushing must be done more often than if the water has low mineral content.

In some locations, the subterranean rocks are so hot that the water from a geothermal power plant vaporizes on its way up through the production well. In this case, the steam can be used to directly drive the turbine. The flash tank is not necessary in this type of system, which is known as a *dry-steam geothermal power plant*.

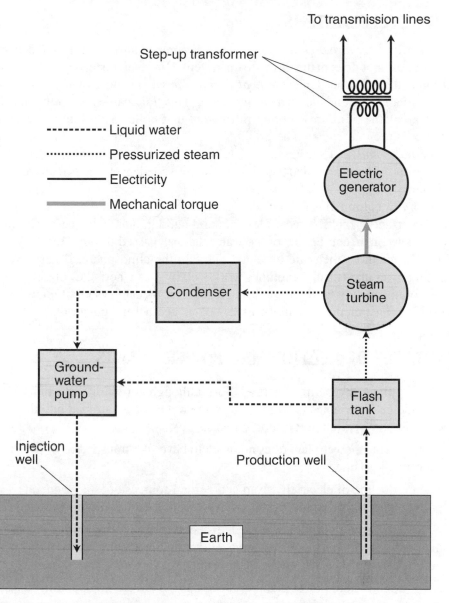

Figure 13-1   Simplified functional diagram of a flash-steam geothermal power plant.

BINARY-CYCLE SYSTEMS

In a *binary-cycle geothermal power plant*, water is pumped into the earth and comes back up hot, just as it does in the flash-steam system. However, instead of going into a flash tank, the hot water enters a *heat exchanger* where most of its energy is transferred to another fluid called a *binary liquid*. This fluid can be water, but more often it is a volatile liquid resembling refrigerant that boils easily into vapor at a lower temperature than water. The liquid-to-vapor conversion occurs in a special low-temperature *boiler*. The pressurized vapor drives a steam turbine. Then the vapor leaves the turbine, is cooled back into liquid by a condenser, and is recirculated to the boiler.

The binary liquid remains in a closed system, isolated from the water that goes into the subterranean rocks. There is less mineral buildup than is the case with the flash-steam system, because none of the water that has passed through the rocks is boiled off. In addition, there are no emissions into the atmosphere. Binary-cycle geothermal power plants can sometimes work well in sites where the subterranean rocks are not hot enough to operate a flash-steam or dry-steam system. Figure 13-2 is a simplified functional diagram of a binary-cycle geothermal power plant.

ADVANTAGES OF GEOTHERMAL POWER PLANTS

- The supply of geothermal energy is vast, although not infinite. It can be considered renewable, as long as excessive water is not pumped into the earth in one location in too short a time.

- A geothermal power plant does not need to have fuel transported or piped in from an outside source.

- The production of electricity from geothermal sources does not generate pollutants or toxic byproducts. (However, see the third limitation below.)

- No external source of fuel is needed, except that required to initially start the pump(s). Once the power plant is operating, the electricity for the pumps can be derived from the plant itself.

- After a geothermal power plant has been built, there are no operating costs, except for routine maintenance and repair.

- A geothermal power plant has a low profile and does not take up a large amount of surface real estate.

- A flash-steam geothermal power plant, if placed on the shoreline of an ocean, can be used to desalinate seawater for drinking and irrigation. This is a natural result of the distillation that occurs when the water is boiled to vapor.

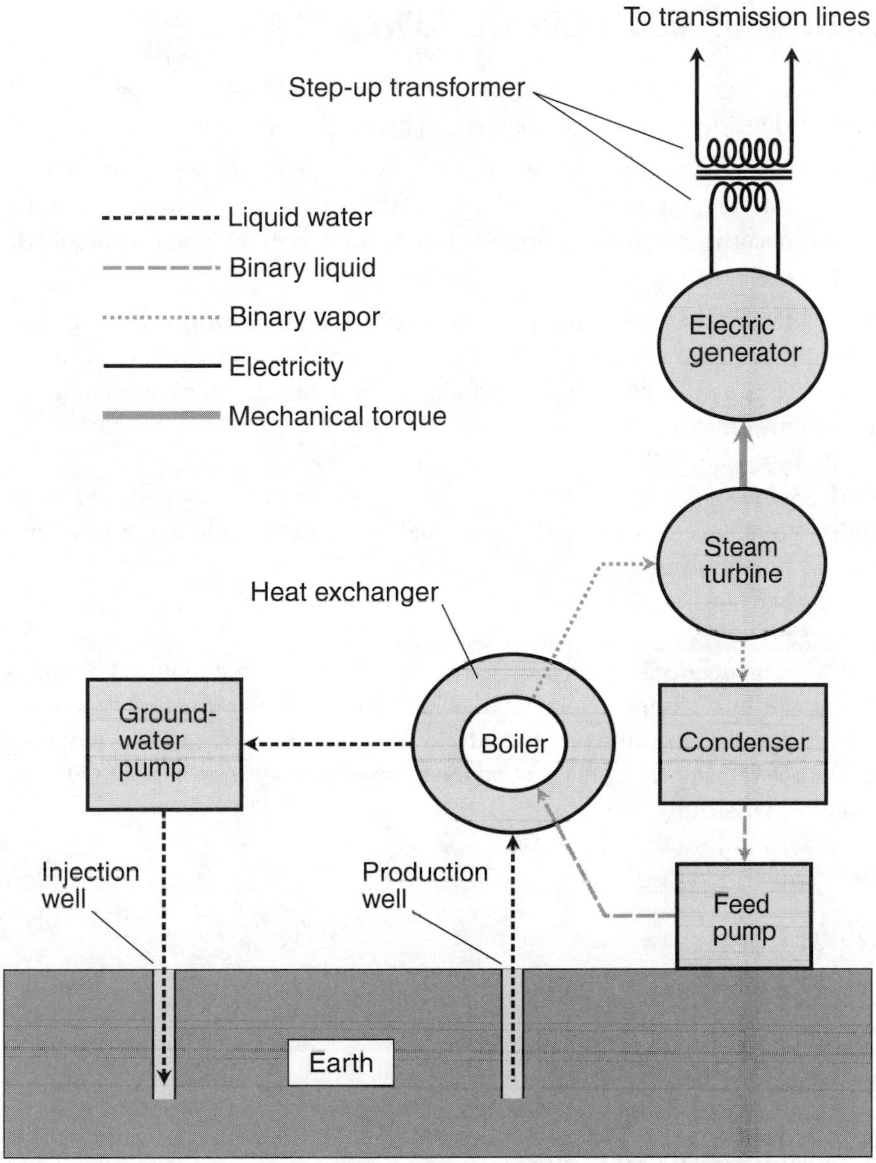

Figure 13-2 Simplified functional diagram of a binary-cycle geothermal power plant.

LIMITATIONS OF GEOTHERMAL POWER PLANTS

- Finding a good site for a geothermal power plant, and getting approval from local residents or governments, can be a big challenge.

- In some cases an established geothermal power plant will "run cold." This can occur as a result of natural changes in the subterranean environment. It can also occur if the site was poorly chosen and too much water is pumped down into the rocks.

- Flammable or toxic gases and minerals may be released from subterranean rocks and come up from the wells. These can be difficult to get rid of. In some cases they can be siphoned off and refined into fuel (crude oil and natural gas, for example).

PROBLEM 13-1

Can a geothermal power plant be built on a small scale to provide electricity for a single home or neighborhood?

SOLUTION 13-1

This can be done in locations where expensive, deep wells are not required. Perhaps the most noteworthy example is Iceland, which in effect sits on top of a huge volcano. Other possible locations include the areas near Yellowstone, Thermopolis, and Saratoga in Wyoming. The area around the town of Hot Springs in South Dakota presents another possibility.

Biomass Power

The term *biomass* describes a wide variety of plant and animal wastes. It literally means "biological matter." Biomass is the oldest source of energy used by humankind, and dates back to the discovery of fire.

BIOMASS ENERGY SOURCES

Biomass is a renewable form of energy because it derives from the *photosynthesis* process, in which plants convert radiant energy from the sun into carbon-containing compounds known as *carbohydrates*. Plants, when grown specifically for use as biomass, actually constitute a form of storage mechanism for solar energy.

When carbohydrates are burned, they release heat, CO_2, and water. The CO_2 goes back into the environment and contributes to the *carbon cycle*, facilitating the growth of more plants to replace the biomass that was burned. Therefore, biomass can be CO_2-neutral if it is responsibly done according to a carefully managed program. The water goes back into the hydrological cycle. The heat energy can be used for electric power generation, as well as for other human energy needs.

Some biomass, such as wood, can be burned straightaway to release its energy. However, various technologies have been developed that allow liquid and gaseous fuels to be derived from wood and other biomass substance. These fuels can be used to supplement (and eventually perhaps replace) gasoline, petroleum diesel, methane, and propane. The following are common raw materials for biomass energy systems:

- **Trees and grasses**   Wood and grass can be directly burned to provide heat for boilers, which drive steam turbines. The most common source of wood biomass is the waste from lumber and paper mills. Willow trees, switchgrass, and elephant grass are grown especially as biomass for energy production.

- **Crops and crop residues**   Corn is used to make ethanol. The same is true to a lesser extent for grains such as wheat, rye, and rice. Sugar cane is used in Brazil to produce ethanol. Soybeans, peanuts, and sunflower seeds have been used to make biodiesel fuel. Both of these fuels, ethanol and biodiesel, can be used for electric power generation, as well as in motor vehicles.

- **Aquatic and marine plants**   Microalgae, found in certain lakes, can be fermented to obtain ethanol or composted to obtain biogas. Ordinary seaweed can also be used for this purpose.

- **Manure and sewage**   Animal waste from farms and ranches, and also human waste from urban areas, can be added to compost piles to accelerate the generation of biogas.

- **Landfills**   Many types of ordinary garbage, particularly paper, cardboard, and discarded food products, can be composted to obtain biogas.

A BIOGAS EXAMPLE

The composting of plant and animal waste can produce combustible methane. Have you heard of the "swamp gas" that can accumulate in wetlands and occasionally catch fire? That's natural biogas! It is essentially the same as commercially or privately produced biogas that can be used for heating, propulsion, and electrification.

Figure 13-3 is a flowchart that illustrates how methane gas can be produced by the composting of plant and animal waste in dedicated facilities for electric power

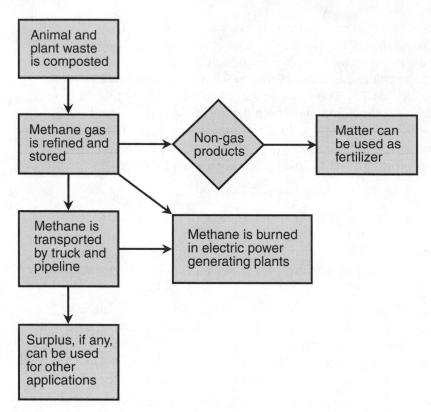

Figure 13-3   Methane gas can be produced by the composting of plant and animal waste in dedicated facilities.

generation and other purposes. Figure 13-4 is a functional diagram of a combined-cycle methane-fired power plant that derives its fuel from the composting of biomass onsite.

ADVANTAGES OF BIOMASS POWER PLANTS

- Biomass is a renewable energy resource.
- Biomass power, if responsibly used, produces zero net CO_2 emissions because the new fuel grown absorbs all the CO_2 generated by the fuel burned.
- Biomass fuel does not produce very much sulfur-based pollution (SO_x), even when it is directly burned. In general, the SO_x production is less with biomass fuels than with conventional fossil fuels.

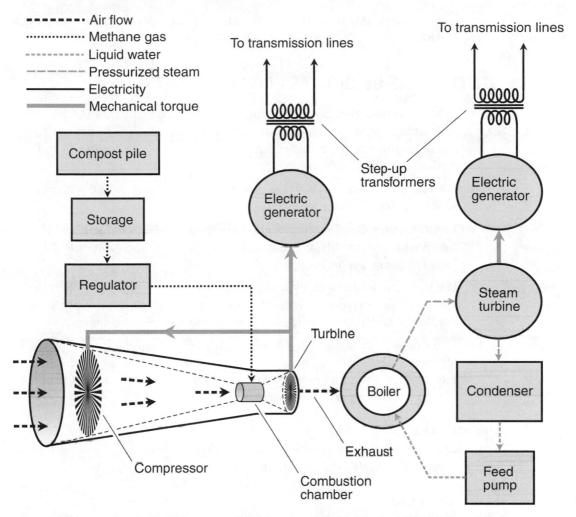

Air flow
Methane gas
Liquid water
Pressurized steam
Electricity
Mechanical torque

To transmission lines

To transmission lines

Compost pile

Storage

Regulator

Electric generator

Step-up transformers

Electric generator

Steam turbine

Turbine

Compressor

Combustion chamber

Boiler

Exhaust

Condenser

Feed pump

Figure 13-4   A combined-cycle methane-fired power plant that derives its fuel from the composting of biomass onsite.

- Large biomass power plants can operate on a continuous basis, unlike solar and wind power plants that produce energy only when the sun shines or the wind blows.

- Methane can be produced in small-scale composting plants. The supply does not have to come exclusively from centralized sources. This could enhance the security of the civilized world by distributing energy resources and assets, making them less vulnerable to natural or human-caused disasters.

- Some of the plants used for biomass power, such as switchgrass, can reduce erosion and provide a habitat for wildlife.

LIMITATIONS OF BIOMASS POWER PLANTS

- Biomass combustion generates some pollutants. The nature of the pollutants depends on the fuel burned. Nitrogen oxides (NO_x) are fairly common. Burning plant matter directly can generate significant CO and particulate pollution.

- Collection of matter for biomass power plants can impact the environment in adverse ways if not responsibly done.

- The transportation of biomatter to composting plants, or to facilities where they are burned directly, consumes energy, usually in the form of fossil fuels for trucks and trains.

- The production of biogas by composting can produce objectionable odors. There is also some concern that the process, if not done responsibly, could lead to the breeding and spread of disease-causing microorganisms.

- Tanks or other containers that hold biogas require periodic inspection and certification by licensed and qualified personnel. This can be inconvenient and costly, but it is an absolute requirement to ensure the safety of people who live and work near the system.

PROBLEM 13-2
Can biogas be derived from small-scale composting and stored onsite for the electrification of a single home or neighborhood?

SOLUTION 13-2
This has been done using backyard compost heaps, makeshift storage containers, and small methane-powered electric generators. However, there can be odor problems. A more serious concern is the onsite storage of methane gas, which can present a fire or explosion hazard. Before such a system is set up, local zoning and fire ordinances should be checked. After the system has been built, it should be initially and periodically inspected by a qualified civil engineer to be sure that it is safe and remains safe.

Small-Scale Fuel-Cell Power

A fuel cell converts combustible gaseous or liquid fuel into usable electricity, but at a lower temperature than normal combustion. In practice, a fuel cell behaves like a battery that can be recharged by filling a fuel tank, or if the fuel is piped in, by a continuous external supply. In Chapter 8 you learned the basics of fuel-cell operation, and how fuel cells can be used for propulsion. The same technology can be used for small-scale electrification.

HOW IT WORKS

Figure 13-5 is a functional block diagram of a small-scale fuel-cell power plant suitable for use in a home or small business. This system can also work for recreational vehicles (RVs) and boats. In the case of a fixed land location, the fuel can be either

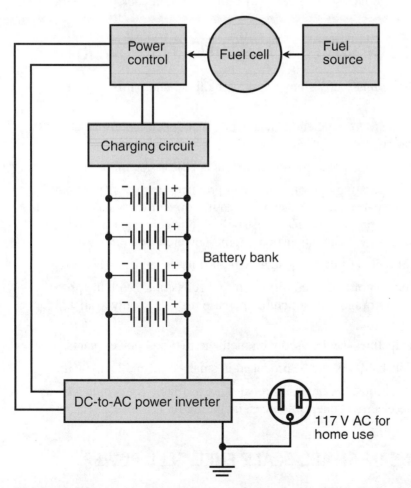

Figure 13-5   Functional block diagram of a small-scale fuel-cell electric power plant, suitable for use in a single residence or small business.

stored onsite or piped in. Conventional methane has been suggested as the ideal fuel source for home power plants of this type because the delivery infrastructure is already in place, and onsite storage is not necessary. However, in rural areas, or in any location not served by methane pipelines, other fuels might prove more cost effective.

A typical fuel-cell stack delivers several volts DC, comparable to the voltage produced by a solar array or car battery. Under normal conditions, the DC from the fuel cell goes to a power inverter that produces usable 117 V AC output. If desired, a backup battery bank can be employed to keep the electric current flowing when the fuel tank is refilled. A power control system switches the electrical appliances between the fuel cell and the battery bank as necessary. The DC output from the fuel cell or battery bank can also be used directly to power small appliances designed to run on low-voltage DC, such as two-way radios and notebook computers.

ADVANTAGES OF SMALL-SCALE FUEL-CELL POWER

- Fuel cells are inherently simple, last a long time, and rarely require maintenance.
- Fuel cells are more efficient than conventional generators for small-scale electrification.
- Hydrogen, a favored energy source for fuel cells, is nontoxic.
- A properly operating hydrogen fuel cell produces negligible pollutant gases and no particulate emissions. Even if more conventional fuels such as methane or propane are used, less pollution is generated with a fuel cell than is the case with a combustion-engine-powered generator.
- The use of fuel cells can help society reduce its dependency on foreign oil.
- The production of hydrogen for use in fuel cells could, with the proper delivery and storage infrastructure, increase the fuel supply available for heating.
- Existing pipelines can be used with methane fuel-cell power plants.
- Some workable fuels can be produced in small-scale and local facilities.
- In some locations, tax breaks or rebates are available for people who use alternative electrical energy sources, including fuel cells.

LIMITATIONS OF SMALL-SCALE FUEL-CELL POWER

- In some areas, it is difficult to get fuel cells serviced because of a lack of parts or competent technicians.

- The delivery and storage of fuel for hydrogen fuel cells presents a major technological obstacle to the widespread deployment of small-scale power plants of this type. (This is not a problem with liquid alternative fuels.)

- The energy density of hydrogen is relatively low compared with other fuels. (This is not a problem with liquid alternative fuels.)

- Hydrogen is extremely flammable and potentially explosive. (This is not usually a problem with liquid alternative fuels.)

- Hydrogen fuel cells are relatively expensive to operate, largely because of the cost of the processes involved in separating hydrogen from naturally occurring compounds. (This is not usually a problem with liquid alternative fuels.)

- Some fuels, such as petrodiesel and biodiesel, tend to solidify in cold weather. This could render a fuel cell inoperative.

- Some fuels, notably methanol and gasoline, can be toxic to personnel directly exposed to them.

PROBLEM 13-3

Can a system such as the one shown in Figure 13-5 be expanded to take advantage of other sources of energy, such as a solar array or a wind turbine, or both? How might this be done? Could that allow a home to operate entirely off the electric utility grid?

SOLUTION 13-3

This is possible but expensive. For example, a solar panel or array can charge the battery bank during sunny weather. A wind turbine can supplement this, taking over on windy nights or windy, overcast days. The fuel cell can operate when neither wind nor solar energy is sufficient to meet the electrical needs of the home or business. A computer-governed power-control switch can ensure that the available energy is used in the most efficient manner at all times. Such a hybrid system can offer complete independence from the electric utility. The key is diversity and redundancy of energy sources.

Aeroelectric Power

Let's end this book by letting our imaginations run a little bit wild. Some scientists think that atmospheric electricity can be tapped to get usable energy. We can go all the way back to Benjamin Franklin (and some less lucky colleagues) who lofted electrical conductors into the air and demonstrated that clouds are electrified.

Suppose some imaginative group of engineers repeats this experiment on a grander scale?

THE GLOBAL ELECTRIC CIRCUIT

The earth is a fairly good *electrical conductor*. So is the upper part of the atmosphere known as the *ionosphere*. The lower atmosphere does not normally conduct electricity, so it composes an *electrical insulator*. When an insulator is sandwiched between two conductors, that insulator is known as a *dielectric*, and the result is a *capacitor* capable of storing energy as an *electric field*. On a gigantic scale it can be called a *supercapacitor*. The *earth-ionosphere supercapacitor* is constantly charging up in some regions and discharging in others, forming a system that has been termed the *global electric circuit*. If it ever becomes practical to tap the global electric circuit to get usable electricity, we will have an *aeroelectric power plant*.

HOW MUCH POWER?

The maximum *electrostatic charge quantity* (number of charged particles such as electrons) that a capacitor can hold depends entirely on three factors: the combined areas of the conducting surfaces, the average distance separating them, and the type of dielectric material between them. The earth-ionosphere supercapacitor consists of one vast conducting sphere inside another with air as the dielectric, as shown in Figure 13-6. These spheres are both about 6500 km (4000 mi) in radius. This forms a capacitor with two "plates" whose spacing is small (about 50 km or 30 mi) compared with their surface areas (about 530,000,000 km² or 200,000,000 mi²).

A high voltage between the earth's surface and the ionosphere gives rise to a massive electric field in the troposphere and stratosphere. The charge in this supercapacitor is maintained by radiation from the sun, from cosmic rays, and from radioactivity in the earth's crust. All of this radiation interacts with the earth's magnetic field and with atoms in the upper atmosphere to keep the supercapacitor charged up.

Storm clouds, volcanoes, and dust storms tend to improve the local conductivity of the troposphere and stratosphere, creating attractive environments for electrical discharge of the earth-ionosphere supercapacitor. A typical thundershower discharges about 2 A of current, averaged over time. At any given moment, there are about 750 thundershowers in progress on our planet, producing between 35 and 100 lightning discharges per second altogether. A current of 2 A per thunderstorm may seem small, but this current does not flow continuously. It occurs in brief, intense surges. A single lightning discharge lasts only a few thousandths of a second. Therefore, the *peak* current in a lightning stroke is extremely large—in some cases many thousands of amperes. This is why lightning can be destructive.

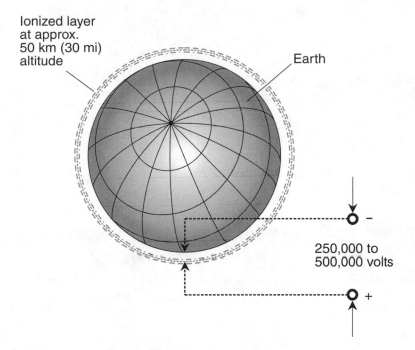

Ionized layer
at approx.
50 km (30 mi)
altitude

Earth

250,000 to
500,000 volts

−

+

Figure 13-6   The earth and the upper atmosphere act as a supercapacitor that is
constantly recharged by various sources of radiation.

The atmospheric supercapacitor maintains a constant charge of 250,000 to
500,000 V, comparable to the voltage in high-tension utility lines. But the earth-
ionosphere electrical potential difference is a DC voltage, not an AC voltage. The
average current that flows across the atmospheric capacitor as a result of
thundershowers alone is roughly 1500 A (2 A per storm times 750 storms). Electrical
power in watts is the product of the voltage in volts and the current in amperes. The
above figures mean that our atmosphere is constantly dissipating several hundred
million watts of power on the average, enough to provide all the electricity needed
by a medium-sized city at peak demand.

AN AEROELECTRIC POWER PLANT

How might an aeroelectric power plant work? One approach would involve lofting
a set of captive high-altitude balloons tethered by conducting wires. The wires
would be grounded through tanks, each containing a solution of water and dissolved
electrolyte (see Figure 13-7). If such a balloon is high enough above the surface to
reach into the lowest ionized layer of the atmosphere, a constant electric current

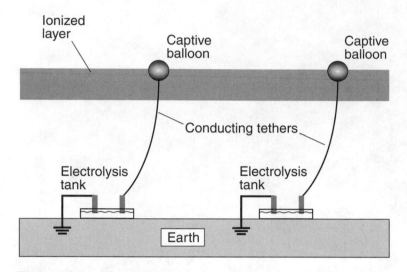

Figure 13-7   A possible scheme for using atmospheric electricity to derive hydrogen fuel from water by electrolysis.

will flow in the wire, and therefore through the electrolyte solution. This will separate the water into hydrogen and oxygen gas, which will bubble from the electrodes. The gases could be collected in the same manner as with any other electrolysis device. The hydrogen could be used in fuel cells or hydrogen-powered cars and trucks. The oxygen could be used for industrial and medical purposes.

ADVANTAGES OF AEROELECTRIC POWER PLANTS

- The earth-ionosphere supercapacitor is constantly recharged by renewable energy sources, notably the sun and various radiation-producing elements in the earth.

- An aeroelectric power plant would produce no pollutants of any kind.

- The facilities for an aeroelectric power plant would be unobtrusive. The balloons would be too far aloft to be seen from the ground without binoculars or telescopes.

- The energy supply could be continuous if captive balloons were kept aloft at all times.

LIMITATIONS OF AEROELECTRIC POWER PLANTS

- Atmospheric electricity, like solar and wind energy, cannot be easily stored. It must be used directly from the source, or else converted to some other form such as hydrogen gas.

- If the earth-ionosphere supercapacitor were discharged to a large extent, it might alter the balance of the global electric circuit. The environmental effects are hard to predict.

- The high voltages in an aeroelectric system could be dangerous to technicians and other personnel working with the equipment.

- Captive balloons of the required size and altitude would be difficult to maintain and keep aloft. They could also present a hazard to aviation.

- The total available energy from atmospheric electricity is limited. It would be, at best, a minor supplement to other energy sources.

PROBLEM 13-4
What would happen to a captive aeroelectric balloon and its associated tether if a heavy thunderstorm were to pass by, as shown in Figure 13-8? (Let's not even think about hurricanes or tornadoes.)

SOLUTION 13-4
The current flowing in the wire and the electrolysis tank would likely change because of localized electric charge poles in the thunderstorm. The air turbulence would stress the wire. If the wire were to snap, the portion on the earth side of the break would fall to the surface near the power plant, and the portion on the balloon side would come down with the balloon at a distant location. This could wreak havoc with the civil infrastructure! If an aeroelectric power plant is ever constructed, it is reasonable to suppose that it will be done on a small island in the middle of a large ocean, and there will be a backup tether or two for the balloon! Don't try it at home.

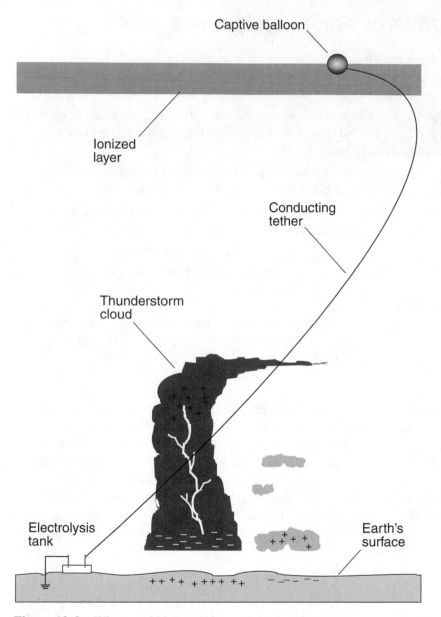

Figure 13-8 What would happen if a captive aeroelectric balloon were to encounter a heavy thunderstorm?

Quiz

This is an "open book" quiz. You may refer to the text in this chapter. A good score is eight correct. Answers are in the back of the book.

1. Which of the following substances is *not* a good source of biomass-derived fuel?

   a. Microalgae

   b. Corn

   c. Sand

   d. Sugar cane

2. The potential difference between the earth and the ionized layers of the atmosphere, for purposes of aeroelectric power generation, is on the order of

   a. a few volts.

   b. a few hundred volts.

   c. a few thousand volts.

   d. a few hundred thousand volts.

3. How deep must the wells be sunk in order for a flash-steam geothermal power plant to be effective?

   a. Deep enough to reach subterranean water in an actively boiling state

   b. Deep enough so the temperature reaches or exceeds the boiling point of water

   c. Deep enough to reach magma that flows freely

   d. Deep enough to reach steam that rises to the surface under its own pressure

4. A thunderstorm produces about the same current, averaged over time, as a 250-W light bulb draws from the electric utility. But the *peak* current in a lightning stroke is much higher than the current drawn by a 250-W light bulb. Why?

   a. The voltage between the storm cloud and the ground is far greater than the voltage provided by the electric utility.

   b. Individual lightning discharges are intermittent and of short duration, not continuous like the discharge through a light bulb.

   c. The earth-ionosphere capacitor can hold a much greater charge than any electric utility generator can produce.

   d. All of the above

5. The electrostatic charge quantity that a capacitor can hold depends on all of the following *except*

   a. the voltage that appears between the two conducting surfaces.

   b. the sum of the areas of the two conducting surfaces.

   c. the nature of the material between the two conducting surfaces.

   d. the average distance between the two conducting surfaces.

6. Suppose the temperature of the soil at the surface is 20°C. How deep should we expect to have to dig a well to reach rocks that are at the boiling point of water?

   a. 286 m

   b. 350 m

   c. 2.86 km

   d. 3.50 km

7. In a flash-steam geothermal power plant, the injection well

   a. serves to deliver water into the earth, where it is heated.

   b. allows red-hot magma to rise to the surface.

   c. allows steam to rise to the surface.

   d. dissipates excess heat from the power-generation process.

8. Which of the following is not practical for use as the energy source in a fuel cell?

    a. Methanol

    b. Hydrogen

    c. Helium

    d. Methane

9. A fuel-cell stack (not an individual fuel cell) delivers about the same voltage as

    a. a large electric generator.

    b. a household utility outlet.

    c. a flashlight cell.

    d. an automotive battery.

10. In a dry-steam geothermal power plant,

    a. a flash tank is not needed.

    b. steam is pumped down the injection well.

    c. steam on the way down the injection well drives a turbine directly.

    d. All of the above

Final Exam

Do not refer to the text when taking this exam. A good score is at least 75 correct. Answers are in the back of the book. It's best to have a friend check your score the first time, so you won't memorize the answers if you want to take the exam again.

1. In passive solar heating, longwave infrared radiation

   a. comes from the sun, passes through the window glass, and is changed to shortwave infrared when it is absorbed by light-colored walls and furniture.

   b. comes from the sun, passes through the window glass, and is completely absorbed by light-colored walls and furniture.

   c. helps to heat up a room as it is radiated from dark-colored objects, because the window glass is relatively opaque to it.

   d. comes from dark objects in a room, radiates through the window glass easily, and is forever lost, be it daytime or nighttime.

   e. can heat the windows to such high temperatures that breakage may occur in cold weather unless double-pane or triple-pane glass is used.

2. The gasoline-gallon equivalent (GGE) of propane is approximately 1.5. This means that

 a. 1 gal of gasoline provides approximately 1.5 times the energy, when burned outright, as 1 gal of propane.

 b. 1 gal of propane provides approximately 1.5 times the energy, when burned outright, as 1 gal of gasoline.

 c. 1 gal of gasoline weighs approximately 1.5 times as much as 1 gal of propane.

 d. 1 gal of propane weighs approximately 1.5 times as much as 1 gal of gasoline.

 e. None of the above

3. A large wood stove, when operating properly, can provide approximately

 a. 150 Btu/h.

 b. 1500 Btu/h.

 c. 1.5×10^4 Btu/h.

 d. 1.5×10^5 Btu/h.

 e. 1.5×10^6 Btu/h.

4. The capacity factor of wind energy is

 a. the ratio of the electrical power output produced by a wind turbine to the actual power contained in the wind.

 b. the ratio of the actual power contained in the wind to the electrical power output produced by a wind turbine.

 c. the proportion of time, for a given location, that the wind can be exploited to get usable electricity from a wind turbine.

 d. the function defining the actual power contained in the wind versus the wind speed in meters per second.

 e. the ratio of the maximum wind speed (in knots) at which a wind turbine will operate properly to the minimum wind speed (in knots) at which the same turbine will operate properly.

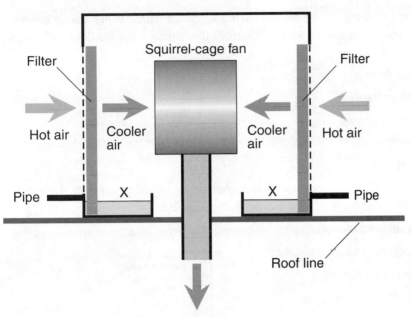

Figure Exam-1 Diagram for questions 5 and 6.

5. Figure Exam-1 is a functional diagram of

  a. an air source heat pump.

  b. a dehumidifier.

  c. an evaporative cooling system.

  d. a passive electrolytic cooling system.

  e. an active electrolytic cooling system.

6. In Figure Exam-1, what substance is represented by the symbols X?

  a. Refrigerant

  b. Ethanol

  c. Methanol

  d. Propane

  e. Water

7. Fill in the blank to make the following sentence true: "When an air source heat pump operates to cool the interior environment, the outdoor air serves as a _____."

   a. heat source

   b. compressor

   c. fluorocarbon

   d. condenser

   e. heat sink

8. The potential energy contained in a specific parcel of water (say, 10 liters), stored in a reservoir behind a dam, is expressed in

   a. meters per second.

   b. meters per second squared.

   c. kilogram-meters.

   d. newton-meters.

   e. newtons.

9. Which of the following technologies has been proven effective for the storage of wind-generated energy in a *large-scale* system for later use?

   a. Composting

   b. Storage batteries

   c. Ultracapacitors

   d. Ground currents

   e. None of the above

10. Fill in the blank to make the following sentence true: "In a northerly place such as Minnesota, the gasoline produced for use in the winter _____ than the gasoline produced for use in the summer."

    a. is more volatile

    b. is more dilute

    c. contains more carbon

    d. is oilier

    e. contains more sugar

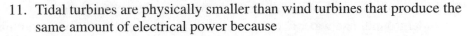

11. Tidal turbines are physically smaller than wind turbines that produce the same amount of electrical power because

    a. ocean currents generally flow more slowly than the wind blows.

    b. water is more dense than air, so it exerts greater force per unit area.

    c. ocean currents tend to flow in a constant direction, but the wind direction varies.

    d. ocean currents tend to flow with constant speed, but the wind speed varies.

    e. Forget it! Tidal turbines are not smaller than wind turbines, but larger.

12. A silicon PV panel provides

    a. output voltage of approximately 0.5 V in direct sunlight.

    b. maximum deliverable current of approximately 2.0 A in direct sunlight.

    c. maximum power output of approximately 1 W in direct sunlight.

    d. either DC or AC, depending on how the cells are connected together.

    e. output voltage, current, and power that depend on numerous factors.

13. Which of the following represents a serious limitation of *large-scale* PV systems for generating electricity?

    a. The PV cells contain cadmium, a toxic heavy metal. Over time this substance "leaches out" of the cells, contaminating the soil and ground water.

    b. The energy produced by PV cells cannot be stored easily on a large scale.

    c. Large photovoltaic arrays emit silicon dioxide, a toxic gas, when exposed for long periods to bright sunlight.

    d. Solar energy is not a renewable resource.

    e. None of the above

14. In an advanced oilheat furnace, the combustion process involves

    a. atomizing the oil and mixing the oil droplets with water vapor.

    b. atomizing the oil and mixing the oil droplets with air.

    c. atomizing the oil and mixing the oil droplets with methane gas.

    d. atomizing the oil and mixing the oil droplets with propane gas.

    e. burning the oil in a pan contained in a firebox.

15. Imagine a power plant that operates by pumping water into hot rocks underground. The water rises back up, turning to steam on the way, and this steam drives a turbine to generate electricity. What type of power plant is this?

   a. A dual-cycle geothermal power plant

   b. A coolant-based geothermal power plant

   c. A dry-steam geothermal power plant

   d. A flash-steam geothermal power plant

   e. A diversity type geothermal power plant

16. What effect does a superconductor have on magnetic lines of flux?

   a. Dilation

   b. Concentration

   c. Bending

   d. Reflection

   e. No effect

17. In order to achieve magnetic levitation with a set of permanent magnets,

   a. some of them must move or rotate relative to the others.

   b. all of them must be stationary relative to the others.

   c. some of them must be aligned vertically.

   d. all of them must be aligned at right angles.

   e. some of them must be aligned horizontally.

18. What type of geothermal power plant can sometimes work where the subterranean rocks are not hot enough to operate a flash-steam or dry-steam system?

   a. Dry-ice

   b. Single-well

   c. Reverse-osmosis

   d. Binary-cycle

   e. Direct-magma

19. Which type of electrical power plant most readily lends itself to the desalination of seawater as a secondary function?

    a. Aeroelectric

    b. Photovoltaic

    c. Flash-steam geothermal

    d. Combined-cycle methane

    e. Hydroelectric

20. The British thermal unit per hour (Btu/h) is a unit of

    a. energy.

    b. heat.

    c. power.

    d. torque.

    e. electrical charge.

21. A typical electric vehicle (EV) gets its power from

    a. a hydrogen battery.

    b. a lead-acid battery.

    c. a mercury battery.

    d. an alkaline battery.

    e. a nitrogen battery.

22. Consider three hydrogen atoms. The first atom has a nucleus consisting of a single proton and nothing else. The second atom has a nucleus consisting of one proton and one neutron. The third atom has a nucleus consisting of one proton and two neutrons. These three atoms represent different

    a. ions of hydrogen.

    b. phases of hydrogen.

    c. atomic numbers of hydrogen.

    d. elements of hydrogen.

    e. isotopes of hydrogen.

23. A hydrogen atom whose nucleus has one proton and two neutrons is known
    as

    a. dilithium.

    b. deuterium.

    c. tritium.

    d. trilithium.

    e. neutrite.

24. In a passive solar heating system that uses large, south-facing windows to
    let sunlight into a house, the most significant source of furniture and carpet
    fading is

    a. ultraviolet radiation.

    b. visible light.

    c. shortwave infrared radiation.

    d. longwave infrared radiation.

    e. the loss of heat energy during hours of darkness.

25. A geothermal power plant might "run cold" if

    a. salt water, rather than fresh water, is pumped down into the
       subterranean environment.

    b. too much water is pumped into the earth at a single location in a short
       period of time.

    c. a volcano erupts nearby, releasing all of the subterranean heat and
       magma pressure.

    d. underground radioactive materials decay beyond their half lives,
       thereby cutting off the source of interior earth heating.

    e. Forget it! A geothermal power plant can never run cold.

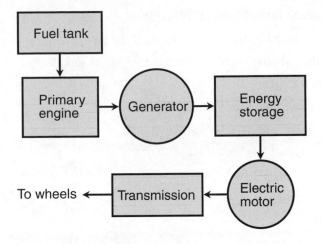

Figure Exam-2 Diagram for question 26.

26. Figure Exam-2 is a block diagram of

  a. a fuel-cell vehicle.

  b. an electric vehicle.

  c. a hybrid electric vehicle.

  d. a photoelectric vehicle.

  e. None of the above

27. Fill in the blank in the following sentence to make it true: "The _____ of an element is approximately equal to the sum of the number of protons and the number of neutrons in the nucleus."

  a. atomic weight

  b. atomic number

  c. isotope number

  d. critical mass

  e. tokamak number

28. A jet aircraft engine produces forward thrust
    a. only when the exhaust speed is higher than the forward airspeed.
    b. only if the speed of the exhaust is greater than the speed of sound.
    c. regardless of the exhaust speed.
    d. only when operated above a certain altitude.
    e. only when operated above a certain forward airspeed.

29. When added to gasoline, ethanol has the effect of
    a. increasing the mileage.
    b. increasing the octane rating.
    c. degrading the performance.
    d. causing more CO gas emissions.
    e. causing more particulate emissions.

30. Magnetic confinement is a technique that has been tested for use in prototypes of
    a. electrostatic power generators.
    b. fusion power generators.
    c. ionospheric power generators.
    d. electrolytic power generators.
    e. geomagnetic power generators.

31. The theoretical upper limit to the speed that an ion rocket can achieve in interstellar space is
    a. less than the theoretical upper limit to the speed that a conventional rocket can attain in interstellar space.
    b. equal to the speed at which the particles are ejected from the engine.
    c. approximately the same as the speed required to escape from the earth's gravitational pull.
    d. the same as the speed required to put a spacecraft into earth orbit.
    e. None of the above

32. In a combustion-turbine type oil-fired electric generator, the turbine itself is driven by

    a. the exhaust from the combustion chamber.

    b. hot water from a boiler.

    c. steam from a boiler.

    d. water pumped from a lake, river, or ocean.

    e. compressed oil from the oil pump.

33. In a forced-air heating system, the amount of dust in the house can be minimized by

    a. using an indoor intake vent rather than an outdoor one.

    b. using an outdoor intake vent rather than an indoor one.

    c. running the fan continuously.

    d. opening all the doors inside the house to optimize air circulation.

    e. using a dehumidifier.

34. The solid fuel in a model-rocket engine bears a strong resemblance to

    a. paraffin.

    b. water ice.

    c. frozen oxygen.

    d. frozen hydrogen.

    e. gunpowder.

35. Which of the following is a significant limitation of sidehill construction in the design of a home?

    a. It cannot provide adequate heat in the winter.

    b. It is workable only on the north slopes of hills.

    c. The roof can present a liability problem.

    d. Excessive heat is lost to the earth in cold weather.

    e. All of the above

36. Animal waste from farms and ranches can be added to compost heaps to form a combustible fuel known as

   a. CO_2.

   b. NO_x.

   c. biogas.

   d. U-235.

   e. petrodiesel.

37. In Figure Exam-3, the dark gray box with the white X represents

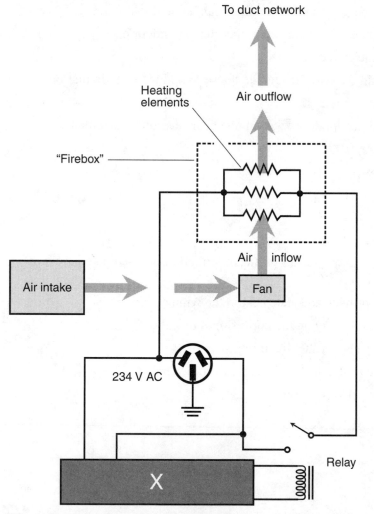

Figure Exam-3   Diagram for question 37.

a. a compressor for heat-transfer fluid.

b. a heat exchanger.

c. an electric voltage regulator.

d. a utility transient suppressor.

e. an electrically powered thermostat.

38. In qualitative terms, standard temperature and pressure are based on

a. the boiling point of water and the atmospheric pressure at sea level.

b. the boiling point of water and the atmospheric pressure at the top of Mount Everest.

c. the freezing point of water and the atmospheric pressure at sea level.

d. the freezing point of water and the atmospheric pressure at the top of Mount Everest.

e. absolute zero and the atmospheric pressure at sea level.

39. Fill in the blank in the following sentence to make it true: "Biomass energy ultimately derives from _____, a process in which plants convert radiant solar energy into carbohydrates."

a. fermentation

b. thermal conversion

c. electrolysis

d. photovoltaics

e. photosynthesis

40. Methane derived from the deliberate, controlled composting of organic matter is called

a. gasohol.

b. swampgas.

c. petrogas.

d. biogas.

e. hydrogas.

41. Fill in the blank to make the following sentence true: "Provided that a cost-effective and efficient means can be devised to extract it from natural sources, and a safe means can be found to transport it, _____ may someday supplement or replace methane as a fuel source."

    a. hydrogen

    b. propane

    c. ethanol

    d. helium

    e. carbon dioxide

42. Residential wind turbines are rarely seen in cities or suburbs mainly because

    a. they can make noise, and some people consider them eyesores.

    b. they kill birds.

    c. there is not enough wind in an urban or suburban environment.

    d. utility companies do not allow wind turbines in incorporated cities.

    e. they attract lightning that could set a town on fire.

43. In a diversion type hydroelectric power plant,

    a. the kinetic energy in moving water is converted to heat, which boils water producing steam; this in turn drives a steam turbine.

    b. a portion of a river is channeled through a canal or pipeline, and the current through this medium drives a water turbine.

    c. water accumulates behind a dam in a large reservoir, and this water is released in a controlled manner through penstocks, driving water turbines.

    d. multiple reservoirs are used at various levels, and water is alternately stored in and released from them by means of pumps and penstocks.

    e. the energy from ocean swells is captured in an enclosed chamber, where varying levels of the water cause air to flow in and out through a gas turbine.

44. Figure Exam-4 is a block diagram of a nuclear fission reactor for use in an oceangoing vessel. What component is represented by the ellipse marked X?

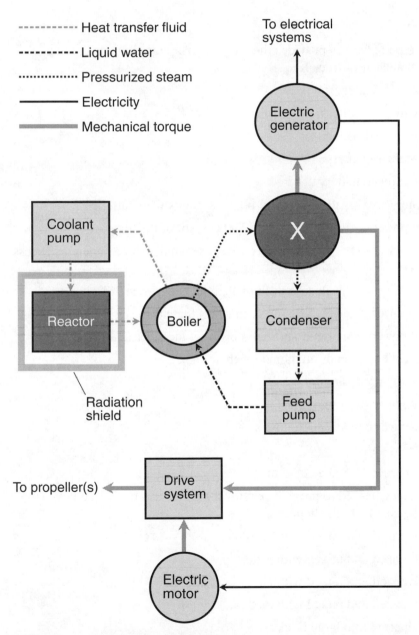

Figure Exam-4   Diagram for question 44.

a. A battery

    b. A turbine

    c. A distiller

    d. A radiator

    e. A motor

45. Which type of battery provides the initial start-up ignition current in a conventional car or truck?

    a. A NiMH battery

    b. An alkaline battery

    c. A lead-acid battery

    d. A mercury battery

    e. A hydrogen battery

46. Most interactive small-scale wind-power systems with batteries

    a. sell power to the electric utility 100 percent of the time.

    b. sell power to the electric utility when the wind turbine generates excess power.

    c. sell power to the electric utility when the batteries are discharging.

    d. sell power to the electric utility in states that allow it.

    e. buy power from, but do not sell power to, the electric utility.

47. Refrigerant compounds are noted for their high

    a. heat of vaporization.

    b. melting or freezing points.

    c. boiling or condensation points.

    d. entropy values.

    e. electrical resistance.

48. Before being burned to provide heat for the boiler in a coal-fired power plant, the coal from the hopper must be

    a. pulverized and mixed with gasoline.

    b. pulverized and mixed with water.

    c. pulverized and mixed with air.

    d. pulverized and mixed with methane.

    e. compacted into large bricks.

49. Which of the following home heating fuels is the most explosive?

    a. Oil

    b. Propane

    c. Butane

    d. Coal

    e. Methane

50. Which of the following is an advantage of a radiant heat subflooring system that uses a methane furnace to heat water?

    a. Deadly CO gas is slow to circulate in the event of a furnace malfunction.

    b. It is a closed system because the water can, in theory, be recirculated indefinitely.

    c. It does not introduce dust into the indoor environment.

    d. The heating coils do not physically intrude into the rooms.

    e. All of the above

51. Which of the following effects or measures, when they occur or are used in conjunction with passive solar heating, can help the interior of a building to remain at a fairly constant temperature despite short-term variations in the solar energy input?

    a. Thermal inertia in concrete floors and walls

    b. The tendency of longwave infrared to pass easily through windows

    c. The use of building materials with as little mass as possible

    d. The use of numerous north-facing windows

    e. All of the above

52. Biomass, if responsibly used, is CO_2-neutral because

    a. the CO_2 resulting from the combustion of biomass combines with hydrogen, forming carbohydrates and water.

    b. the CO_2 resulting from the combustion of biomass re-enters the carbon cycle, facilitating the growth of more plants to replace the spent biomass.

    c. the CO_2 resulting from the combustion of biomass escapes into outer space.

    d. the CO_2 resulting from the combustion of biomass can be electrolyzed, obtaining carbon for heating and oxygen for medical purposes.

    e. Forget it! Biomass can never be CO_2-neutral.

53. When several identical silicon PV cells are connected in series to form a module, the maximum obtainable output power of the module, neglecting internal resistance, is equal to

   a. the maximum obtainable output power of any one of the cells.

   b. the maximum obtainable output power of any one of the cells, multiplied by the square root of the number of cells.

   c. the maximum obtainable output power of any one of the cells, multiplied by the number of cells.

   d. the maximum obtainable output power of any one of the cells, multiplied by the square of the number of cells.

   e. None of the above

54. Carbon nanotubes have been suggested as a method of

   a. improving the efficiency of the water electrolysis process.

   b. producing methane for use in home heating.

   c. storing hydrogen for use as fuel in motor vehicles.

   d. storing electrical energy from solar panels, wind turbines, and water turbines.

   e. storing and releasing electrical energy in regenerative braking.

55. Which of the following is a significant limitation of rooftop flat-plate collectors for the purpose of solar water heating or home interior heating?

   a. They cannot be used with heat-transfer fluid because this fluid has a tendency to freeze in cold weather.

   b. They can be damaged by hail storms or ice buildup, and they must be kept clear of snow in order to function properly.

   c. They will not work at latitudes far from the equator because the sunlight is never intense enough.

   d. They can produce excessive voltage in bright sunlight, and this can result in damage to the storage batteries.

   e. If part of a plate is in shadow and the rest is in direct sunlight, the other panels can be damaged by excessive current.

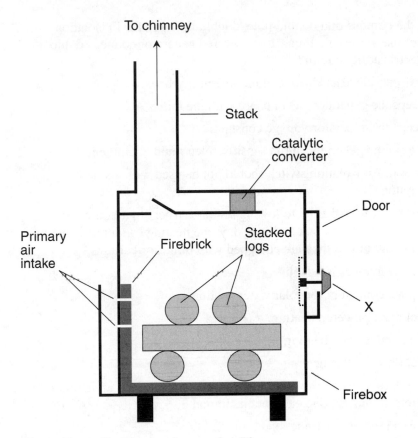

Figure Exam-5 Diagram for question 56.

56. Figure Exam-5 is a cross-sectional diagram of a stove for burning cut wood. What is the function of the component marked X?

    a. It allows control of the operation of the catalytic converter.

    b. It allows control of the rate at which the logs burn.

    c. It prevents explosions in the event of downdrafts in the stack.

    d. It prevents heat from escaping from the firebox by unwanted convection.

    e. It allows excess air from the firebox to escape.

57. What is the purpose of a double-pole, double-throw (DPDT) isolation switch for use with a small gasoline-powered generator connected into a home electric utility circuit?

    a. It prevents electrical backfeed into the utility lines.

    b. It keeps the generator AC in phase with the utility AC.

    c. It keeps the generator voltage constant.

    d. It allows the generator to run sensitive electronic equipment.

    e. Forget it! An isolation switch should not be used with a small generator.

58. Fill in the blank to make the following sentence true: "In a house with _____, the windows that facilitate daytime heating can let heat energy escape at night, unless they are equipped with blinds or curtains."

    a. direct hydroelectric heating

    b. flat-plate collectors for solar water heating

    c. direct wind-powered heating

    d. basic passive solar heating

    e. direct photovoltaic heating

59. The use of E10 instead of regular gasoline in a vehicle with an internal-combustion engine can decrease the likelihood of

    a. electrical system malfunction.

    b. poor mileage because of poor engine timing.

    c. uncontrolled acceleration.

    d. gas-line freeze in cold weather.

    e. All of the above

60. One of the most promising technologies for controlled fusion uses fuel consisting of

    a. deuterium and tritium.

    b. helium and hydrogen.

    c. the most abundant isotope of hydrogen.

    d. hydrogen and oxygen.

    e. hydrogen and nitrogen.

61. Methane is commonly used as a fuel for

    a. jet aircraft.

    b. cars and trucks.

    c. trains.

    d. home heating.

    e. evaporative cooling.

62. What purpose does a transient suppressor serve in a wood pellet stove?

    a. It minimizes the buildup of creosote in the firebox, and improves the overall efficiency of the stove.

    b. It helps to regulate air flow to the firebox, and prevents overheating caused by sudden downdrafts in the flue.

    c. It keeps the pellets burning at a constant temperature, and minimizes the amount of generated ash or "clinkers."

    d. It minimizes the risk of damage to the control electronics in case of a "spike" in the AC utility voltage.

    e. Forget it! A transient suppressor is completely irrelevant to the operation of a wood pellet stove.

63. A disadvantage of methane for electric generation in the United States is the fact that

    a. it produces heavy-metal waste that can take centuries to decay.

    b. it is also a popular home heating fuel, and this can cause supply-and-demand trouble.

    c. it must be shipped by rail or truck from refineries to end users.

    d. it burns inefficiently, producing a lot of pollution and relatively little energy.

    e. it must be pulverized before it can be used in a combustion chamber.

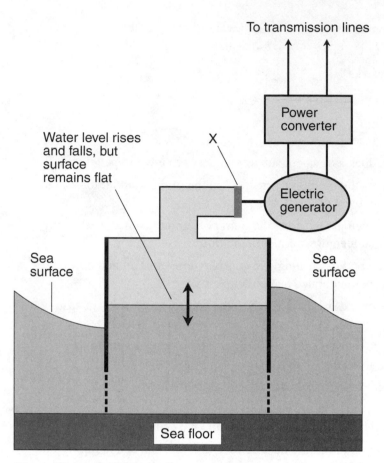

Figure Exam-6 Diagram for question 64.

64. Figure Exam-6 is a simplified functional diagram of a system designed to produce electricity from ocean swells. What type of device should be in the spot marked X?

    a. An air pump

    b. A water pump

    c. An escape valve

    d. An air turbine

    e. None of the above

65. Suppose the resistance of the heating element in a portable electric space heater is doubled. The heater operates at a constant AC voltage of 117 V rms. What happens to the heat power produced by this device, assuming all the electricity in the heating element is converted into usable heat?

    a. It drops to one-quarter of its previous value.

    b. It drops to half of its previous value.

    c. It stays the same.

    d. It doubles.

    e. It quadruples.

66. As a fuel for propulsion, petroleum diesel is known for its

    a. high ethanol content.

    b. high energy density.

    c. high methanol content.

    d. high methane content.

    e. high lead content.

67. When the plant matter used to produce a given quantity of ethanol is replaced by an equal amount of new plant matter in systematic ethanol farming, the ethanol thus produced is

    a. CO_2 enhanced.

    b. CO_2 negative.

    c. CO_2 neutral.

    d. CO_2 sparing.

    e. CO_2 proof.

68. When a diesel engine is under full load (for example, when a heavy truck goes up a steep grade), the exhaust can contain large amounts of elemental

    a. methane.

    b. oxygen.

    c. carbon.

    d. nitrogen.

    e. hydrogen.

69. When an air source heat pump is used for the purpose of warming the interior of a house, the temperature of the air emerging from the indoor fan is approximately

    a. 10°C.

    b. 20°C.

    c. 35°C.

    d. 70°C.

    e. 100°C.

70. In which of the following ways does corn differ from wood pellets for use in home heating?

    a. Corn can be stored indefinitely, but wood pellets cannot.

    b. Corn produces creosote, but wood pellets do not.

    c. Corn contains dangerous biological substances, but wood pellets do not.

    d. Corn produces more ash than wood pellets.

    e. Corn contains ethanol, which burns hotter than wood.

71. Which of the following conditions, all by itself, makes subterranean living impractical?

    a. Frequent tornadoes

    b. Frequent hail storms

    c. Frequent lightning storms

    d. Flat terrain

    e. None of the above

72. A house with a forced-air heating and cooling system should not be completely airtight because if it is,

    a. excessive heat can build up when the furnace runs, causing uncomfortable fluctuations in the temperature.

    b. air conditioning equipment will cause condensation on the upper floors where temperatures are coolest.

    c. air will not be able to circulate freely throughout the house, even if all the interior doors are left open.

    d. CO gas might build up rapidly in the event of a furnace malfunction and reach deadly levels before you can do anything about it.

    e. Hold it! The premise is wrong. A house should be completely airtight to optimize its energy efficiency.

73. When does an evaporative cooling system work best?

    a. When the weather is hot and humid

    b. When the weather is humid and windy

    c. When the weather is warm and cloudy

    d. When the weather is hot and dry

    e. An evaporative cooling system works equally well in any of the above-mentioned weather conditions.

74. Absolute zero represents

    a. the absence of all thermal energy.

    b. the freezing point of hydrogen.

    c. the freezing point of helium.

    d. the freezing point of air at sea level.

    e. the freezing point of oxygen.

75. A coal-only stove (as opposed to a hybrid coal/wood stove) is intended to burn

    a. peat.

    b. anthracite.

    c. bituminous.

    d. lignite.

    e. All of the above

76. Which of the following is an advantage of sidehill home construction compared with conventional construction?

    a. Less exterior maintenance is required.

    b. Such a home stays cooler in summer.

    c. The heat retention is superior if the insulation is good.

    d. Less external noise gets in.

    e. All of the above

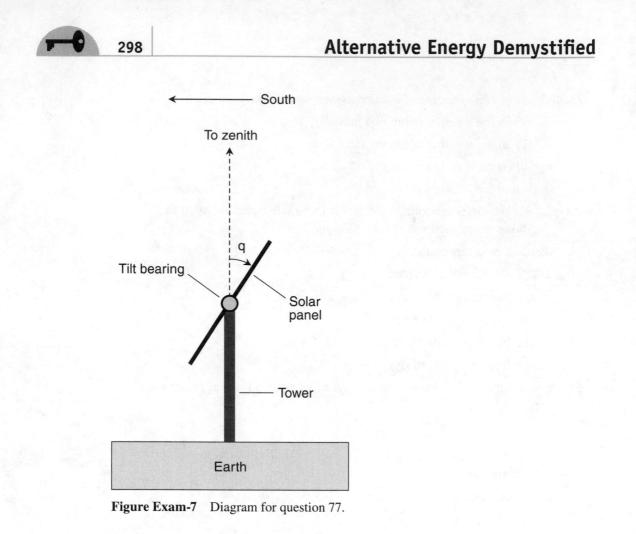

Figure Exam-7   Diagram for question 77.

77. Figure Exam-7 shows a solar panel with an adjustable tilt bearing. The panel is viewed edge-on, looking west. What is the optimum tilt angle q for year-round panel use at 55° north latitude, assuming the bearing will never be adjusted again once it has been set?

  a. 55°

  b. 45°

  c. 35°

  d. 25°

  e. The optimum tilt angle q is impossible to determine without knowing the longitude as well as the latitude.

78. Consider an electric zone heating system with resistive elements in the ceilings. These elements glow red-hot when electric current passes through them. Infrared reflectors are placed above the elements. The primary mode of heating in this system is

    a. radiation.

    b. conduction.

    c. entropy.

    d. convection.

    e. absorption.

79. A rocket engine can produce forward thrust in outer space

    a. only when the exhaust speed is higher than the forward speed of the rocket.

    b. only when the rocket travels faster than a certain minimum forward speed.

    c. regardless of the exhaust speed relative to the rocket.

    d. only when liquid fuel is used.

    e. only when solid fuel is used.

80. Which of the following units expresses the same measurable quantity or phenomenon as the joule?

    a. The degree Celsius

    b. The degree Fahrenheit

    c. The watt per second

    d. The British thermal unit

    e. The calorie per second

81. Fill in the blank in the following sentence to make it true: "The farming of crops for ethanol production takes up _____ from the atmosphere."

    a. oxygen

    b. methane

    c. carbon dioxide

    d. hydrogen sulfide

    e. particulate pollutants

82. Which of the following is a good reason to place the oil tank for an oilheat furnace below ground?

    a. It is easier to repair than an above-ground tank in case it springs a leak.

    b. It is less likely than an above-ground tank to be affected by earthquakes.

    c. The subsurface temperature is more stable than the above-ground temperature, so the oil is less likely to become sluggish in cold weather.

    d. The oil in an underground tank is easier to mix with biofuels than the oil in an above-ground tank.

    e. Forget it! The oil tank for an oilheat furnace should never be below ground.

83. In the United States, biodiesel is derived mainly from

    a. corn.

    b. wheat.

    c. barley.

    d. oats.

    e. None of the above

84. Which of the following proposed technologies would take advantage of the high-speed subatomic particles emanating from the sun in order to propel a spaceship through interplanetary space?

    a. The solar sail

    b. The ion rocket

    c. The fusion engine

    d. The fission engine

    e. The Bussard ramjet

85. Fill in the blank to make the following sentence true: "Most proposals for hydrogen-fusion-powered spacecraft envision using _____ to place the ship in an earth orbit, and then starting up the fusion engine in space."

    a. an ion engine

    b. a conventional rocket

    c. a jet aircraft

    d. a solar sail

    e. a hydrogen bomb with a blast deflector

86. Which of the following is a significant advantage of rooftop flat-plate collectors for the purpose of solar water heating or home interior heating?

    a. They can work well no matter how they are positioned.

    b. They will work well even when covered by snow.

    c. They are virtually indestructible.

    d. They can serve as skylights when placed in a north-facing, steeply pitched roof.

    e. The only necessary external power source is electricity for the fluid pump.

87. Refer to Figure Exam-8. What type of power plant is this?

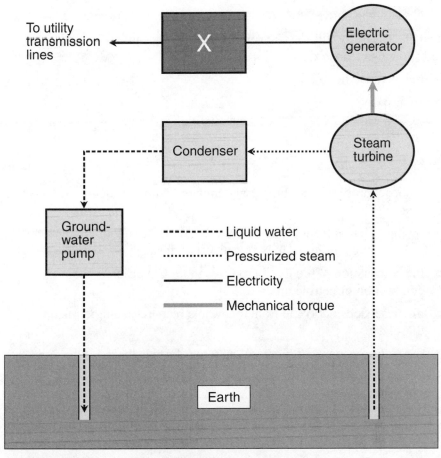

Figure Exam-8   Diagram for questions 87 and 88.

a. A dual-cycle geothermal power plant

b. A coolant-based geothermal power plant

c. A dry-steam geothermal power plant

d. A flash-steam geothermal power plant

e. A diversity type geothermal power plant

88. In the system of Figure Exam-8, what is the component marked X?

a. A power inverter

b. A transient suppressor

c. A battery

d. A transformer

e. None of the above

89. In an electric or hybrid electric vehicle, the battery can be charged by all the following means *except*

a. regenerative braking.

b. solar panels.

c. wind turbines.

d. the electric utility.

e. electrolysis of water.

90. A significant problem with biodiesel compared with petroleum diesel is the fact that

a. biodiesel has solvent properties that can damage rubber components in older engines designed to burn petroleum diesel.

b. the combustion of biodiesel produces more CO gas than the combustion of petroleum diesel.

c. biodiesel does not come from renewable resources, but petroleum diesel does.

d. the widespread use of biodiesel may hurt farmers because it will reduce the demand for certain crops.

e. All of the above

91. A significant advantage of biodiesel combustion over petroleum diesel combustion is the fact that in most situations, biodiesel combustion produces less

    a. carbon dioxide.

    b. sulfur dioxide.

    c. carbon monoxide.

    d. particulate matter.

    e. of all of the above.

92. Which of the following events is most likely to render a small-stream direct hydroelectric home climate-control system inoperative?

    a. An electric utility blackout

    b. A hail storm

    c. A heavy snow

    d. A prolonged drought

    e. All of the above events are equally likely to render a small-stream direct hydroelectric home climate-control system inoperative.

93. Acceleration and deceleration in a maglev train is accomplished by

    a. permanent magnets.

    b. linear motors.

    c. AC electromagnets.

    d. alternators.

    e. Earnshaw effect.

94. Suppose a long 100,000-V transmission line serves a certain small community. A proposal is put forward to increase the voltage of the line to 400,000 V. The new line will be exactly as long as the old one. Wires of the same diameter will be used, so the conductor resistance will be no different. No change is expected in the power used by the community. If the new line is installed, by what factor can the power loss in the line be expected to change?

    a. It will decrease by a factor of 2.

    b. It will decrease by a factor of 4.

    c. It will decrease by a factor of 16.

    d. It will decrease by a factor of 64.

    e. More information is needed to answer this.

95. When energy is allowed to freely move from one medium into another in the form of heat, the temperatures of the two media tend to equalize over time. This process is known as

    a. heat of fusion.

    b. thermal degradation.

    c. thermal distribution.

    d. heat entropy.

    e. potential energy equalization.

96. A modest hydropower plant, designed to generate electricity from a stream or small river for use by a single household, would most likely rely on

    a. diversion technology.

    b. water impoundment.

    c. multiple reservoirs.

    d. tidal barrages.

    e. steam turbines.

97. Steam-turbine oil-fired power plants can be designed so that they can also be fueled by

    a. methane.

    b. uranium.

    c. nitrogen.

    d. nitrous oxide.

    e. sulfur dioxide.

98. Figure Exam-9 is a simplified functional block diagram of a hydrogen fuel-cell vehicle. What is represented by the box marked X?

    a. An ultracapacitor

    b. A carbon nanotube network

    c. An alternator

    d. A regenerative braking system

    e. A storage battery

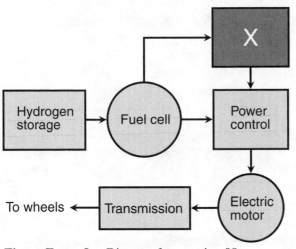

Figure Exam-9   Diagram for question 98.

99. An ultracapacitor can be used in an electric or hybrid electric vehicle to

    a. provide a short burst of acceleration after a period of deceleration.

    b. charge the battery while the vehicle is not in use.

    c. prevent hydrogen and sulfur dioxide emissions from the battery.

    d. cool or heat the interior of the vehicle to keep it comfortable.

    e. change the AC from the generator into DC for charging the battery.

100. What is the function of a power inverter in a direct PV climate-control system?

    a. It reverses the phase of the 117 V AC utility electricity.

    b. It changes the low-voltage DC output of a solar panel into 117 V AC.

    c. It changes the 117 V AC utility electricity into low-voltage DC.

    d. It keeps the output of the PV cells at a constant DC voltage.

    e. Forget it! Direct PV climate-control systems never employ power inverters.

Answers to Quiz and Exam Questions

CHAPTER 1

| 1. a | 2. d | 3. b | 4. c | 5. d |
| 6. c | 7. b | 8. b | 9. a | 10. b |

CHAPTER 2

| 1. c | 2. a | 3. d | 4. a | 5. c |
| 6. b | 7. c | 8. a | 9. b | 10. d |

CHAPTER 3

1. c	2. c	3. b	4. d	5. a
6. d	7. b	8. a	9. b	10. d

CHAPTER 4

1. c	2. c	3. b	4. c	5. a
6. a	7. d	8. b	9. a	10. b

CHAPTER 5

1. c	2. c	3. c	4. b	5. a
6. a	7. b	8. d	9. b	10. b

CHAPTER 6

1. a	2. b	3. c	4. a	5. a
6. c	7. d	8. b	9. c	10. d

CHAPTER 7

1. c	2. c	3. d	4. a	5. a
6. d	7. d	8. c	9. b	10. c

CHAPTER 8

1. c	2. a	3. d	4. d	5. d
6. b	7. b	8. d	9. c	10. b

CHAPTER 9

1. d	2. b	3. d	4. a	5. d
6. d	7. a	8. c	9. b	10. b

CHAPTER 10

1. c	2. a	3. a	4. c	5. a
6. a	7. c	8. d	9. b	10. d

CHAPTER 11

1. a	2. b	3. c	4. d	5. d
6. b	7. d	8. a	9. d	10. c

CHAPTER 12

1. c	2. c	3. a	4. b	5. d
6. b	7. b	8. d	9. c	10. c

CHAPTER 13

1. c	2. d	3. b	4. b	5. a
6. c	7. a	8. c	9. d	10. a

FINAL EXAM

1. c	2. a	3. d	4. c	5. c
6. e	7. e	8. d	9. e	10. a
11. b	12. e	13. b	14. b	15. c
16. a	17. a	18. d	19. c	20. c
21. b	22. e	23. c	24. a	25. b
26. c	27. a	28. a	29. b	30. b
31. e	32. a	33. a	34. e	35. c
36. c	37. e	38. c	39. e	40. d
41. a	42. a	43. b	44. b	45. c
46. e	47. a	48. c	49. e	50. e
51. a	52. b	53. c	54. c	55. b
56. b	57. a	58. d	59. d	60. a
61. d	62. d	63. b	64. d	65. b

66. b	67. c	68. c	69. c	70. e
71. e	72. d	73. d	74. a	75. b
76. e	77. c	78. a	79. c	80. d
81. c	82. c	83. e	84. a	85. b
86. e	87. c	88. d	89. e	90. a
91. e	92. d	93. b	94. c	95. d
96. a	97. a	98. e	99. a	100. b

APPENDIX

Suggested Additional Reading

BOOKS

- Chiras, Daniel. *The Solar House: Passive Heating and Cooling*. White River Junction, VT: Chelsea Green Publishing, 2002.

- Ewing, Rex A. *Power with Nature: Solar and Wind Energy Demystified*. Masonville, CO: PixyJack Press, 2003.

- Gipe, Paul. *Wind Energy Basics: A Guide to Small and Micro Wind Systems*. White River Junction, VT: Chelsea Green Publishing, 1999.

- Kachadorian, James. *The Passive Solar House*. White River Junction, VT: Chelsea Green Publishing, 1997.

- Koomp, Richard. *Practical Photovoltaics: Electricity from Solar Cells*. 3rd ed. Ann Arbor, MI: Aatec Publications, 2001.

- Ritchie, Ralph. *All That's Practical About Wood*. Springfield, OR: Ritchie Unlimited Publications, 1998.

- Strong, Steven, and William Scheller. *The Solar Electric House*. Still River, MA: Sustainability Press, 1993.

COMPACT DISCS

- World Spaceflight News. *21st Century Complete Guide to Biofuels and Bioenergy*. Mount Laurel, NJ: Progressive Management, 2003.

- World Spaceflight News. *21st Century Complete Guide to Geothermal Energy*. Mount Laurel, NJ: Progressive Management, 2004.

- World Spaceflight News. *21st Century Complete Guide to Hydrogen Power and Fuel Cell Cars*. Mount Laurel, NJ: Progressive Management, 2003.

- World Spaceflight News. *21st Century Complete Guide to Nuclear Fusion*. Mount Laurel, NJ: Progressive Management, 2002.

- World Spaceflight News. *21st Century Complete Guide to Solar Energy and Photovoltaics*. Mount Laurel, NJ: Progressive Management, 2004.

- World Spaceflight News. *21st Century Complete Guide to Wind Energy and Wind Turbines*. Mount Laurel, NJ: Progressive Management, 2004.

INDEX